Constitution of Organs of the Higher Plants

Chi Yen

Constitution of Organs of the Higher Plants

The multiple secondary axis theory

中国农业出版社
CHINA AGRICULTURE PRESS

Springer

Chi Yen
Triticeae Research Institute
Sichuan Agricultural University
Chengdu, China

Translated by
Yang Yen
Biology and Microbiology
South Dakota State University
Brookings, SD, USA

ISBN 978-981-19-1684-7 ISBN 978-981-19-1685-4 (eBook)
https://doi.org/10.1007/978-981-19-1685-4

Jointly published with China Agriculture Press
The print edition is not for sale in China (Mainland). Customers from China (Mainland) please order the print book from: China Agriculture Press.

This Springer imprint is published by the registered company Springer Nature Singapore Pte Ltd.
The registered company address is: 152 Beach Road, #21-01/04 Gateway East, Singapore 189721, Singapore

Preface

Agriculture is the foundation of national economy. This is because at present organic matters can only be produced by photosynthesis, which converses energy from light (mainly sunlight) and stores it in organics biosynthesized from inorganics. The organics needed by humans for survival and conducting activities are also produced mainly by crops via photosynthesis. The organics produced by green plants via photosynthesis are the essential resources of the most essential substances needed for life by all organisms including humans. Before humans can imitate photosynthesis using sunlight to produce organics, photosynthesis in green plants that produces organics from inorganics is absolutely the only means to store solar energy. Even when the secret of photosynthesis is completely revealed and artificial photosynthesis could be used to produce organics for storing light energy, agriculture will always be the foundation of national economy; and green plants will still be an important component of the physical world and a part of the material cycle in the eco-environment which humanity is also a part of, and thus is vital to humans for their survival and living. Of course, this is the reality that cannot be changed unless the sun dies, and the solar energy completely disappears.

Crop varieties are the fundamental means of agricultural production. The quality and quantity of agricultural products are directly determined by the economic traits of crop varieties. Therefore, improving crop varieties by breeding is the most important way to increase agricultural productivity. For a long time, selection of new varieties during agricultural production, as Darwin summarized, was unconscious. It had become modern, scientific, and conscious only after the function of artificial selection of improved varieties was recognized. Therefore, spontaneously occurred beneficial genetic variations that met humans' needs were collected, gathered, preserved, selected, and utilized in a relatively short period of time in recent centuries. For instance, some famous varieties of crops, poultries, and livestock bred at that time basically had met the needs by the booming, flourishing, large-scale agricultural production at the early stage of capitalism.

On the other hand, people had recognized, in the practice of artificial selection, that sexual hybridization can produce more variations in the offspring for the benefits of selection. In addition, various good traits can be combined into the offspring through hybridization. A new era of hybridization breeding then began. As a result, genetics was established in the late nineteenth century for the study of the laws of inheritance and variations created by hybridization.

Since the beginning of the twentieth century, due to the advances in genetics, cytogenetics, and molecular genetics, the laws and mechanisms of inheritance and variations of organisms have been understood better and more thoroughly. Development and applications of artificial mutagenesis, chromosome engineering, and DNA engineering and the application of these technologies have started to advance breeding from being just waiting for spontaneous variations and genetic recombinants to occur into a new area of controlling genetic variations. The use of germplasm has been broadened beyond the boundaries of species and genera by wild hybridization to utilize germplasm from related species and by genetic transformation that artificially transfers cloned genes. Now we can create a new crop with brand new traits that have never been seen in nature before by controlling DNA expression and creating new genes via reformatting DNA base pairs. In 2012, approved by USDA and FDA, a new transgenic soybean cultivar Plenish was released by Pioneer Hybrid, which marks the creation of a new soybean cultivar by regulating DNA with transgenic technology. A brand new high-quality plant oil that had not been seen previously in nature has been produced by regulating DNA to suppress the activity of FAD2-1, a key enzyme in oil metabolism in soybean. This new oil contains more oleic acid, but less linoleic acid and linolenic acid, and thus has higher nutrition value and stronger antioxidization activity and is more heat stable. Consumption of this oil can reduce triacylglycerol in the blood; frying food with this oil can greatly reduce carcinogens in the food. Therefore, Plenish is incomparable with any nontransgenic soybean at the time.

From unconscious selection to conscious selection of crop varieties and from hybridization breeding to transgenic breeding, humans have journeyed four stages. The first stage lasted about 10,000 years. The second stage probably started at the beginning of the seventeenth century. Therefore, it has been less than 200 years since humans understood hybridization breeding. Transgenic crops were born in the late twentieth century. Hybridization breeding had led to the understanding of the laws of inheritance and variations, and intensive studies of the fundamentals of genetics has led to the application of transgenic breeding.

On the one hand, modern agricultural production is no longer satisfied with just waiting for random emergence of new types; it demands scientific design of "ideotypes." This requires that breeding evolves from comparative appraisal of the types already in hand with statistical analysis to a new realm of modern scientific breeding that new types are designed to better meet the demands of production. Modern scientific breeding has similar requirements just as other engineering works do. Therefore, breeding new improved crop cultivars this way requires a thorough understanding of the genetic material from molecules to chromosomes, organelles, cells, tissues and/or organs, of the biochemical and physiological pathways they reflect or derive from, and of the laws that govern the evolution of morphological constitution from micro to macro levels and vice versa. Only when we understand all thcsc can we design the new types of cultivars under the guidance of these knowledges, foresee the possibilities of breeding new cultivars and the rationality of their organ constitution and their trait specifications, and can we decide the specific approaches, steps, and technologies used for the creation of such a cultivar. The products of agricultural production are particular organ(s) of the crops, not all organs. However, the product organ is associated with all other organs. After all, improved crops are abnormal plants. The demands by the new designing work for improved new cultivars will never be met if we only understand the initial molecular structures and mechanisms of the genetic materials but not the laws of the organ evolution reflected by the genetic traits, the physiology and biochemistry of the functional processes, and the organogenesis initiated by the genetic materials under the specific environment conditions. From our basic understanding of the evolutionary patterns of organs and without a thorough understanding of this knowledge, it is impossible to foresee the possible organ variations and the potential breeding process, to make effective designing of reasonable organ structures, or to select the related traits predictably and effectively; it will also not be possible to predict the occurrence of new organ traits beforehand or to reach the goal of breading the new cultivars according to the design plan.

Investigating the evolution of plant organ structures is the task of modern plant developmental morphology. Before modern breeding starts to design new crop structures, there is no immediate need for the study of the basic theory of this branch of science because it apparently does not have a close tie with breeding. When breeders start to analyze and design reasonable organ structures for creating a new cultivar, the theory of the organ morphology of higher plants will become a fundamental knowledge of breeding. Unfortunately, when this basic theory is in need of application, we not only find that the current research status of this ancient branch of botany is too far behind the needs, but also identifies conflicts in principles between the breeding practice and the current theories of organogenesis. Practice is the sole standard for testing the correctness of a theory. Practices have revealed many errors in the current theories of the organ morphology of higher plants. Only conducting further research to correct these errors can we provide reliable scientific basis for new cultivar design and more freedom for the designing and analysis. In his practices, the author has reviewed, re-analyzed, and re-investigated questions about the structural unit of organ morphology and the mechanisms of organogenesis in higher plants. On the basis of these studies, a multiple secondary axis theory has been formed to describe the structural unit of organ morphology and the mechanisms in organogenesis in higher plants.

During the "Great Cultural Revolution," the author overcame interferences and completed two big tasks. The first is the improvement of the low-yield, poor-quality wheat in Sichuan by breeding high-yield, high-quality wheat cultivars that were suitable to grow in the Sichuan wheat production region. Guided by my improved convergent hybridization plan, the author released the stripe rust-resistant, high-yield Fan 6/Fan 7 sister lines in 1969 after 5 years of breeding efforts. These cultivars increased the wheat yield in the Sichuan wheat production region from 3000–3750 to 5250–6000 kg ha hm^2 and thus lifted the wheat yield in the region one stage higher. Multiple resistance genes were converged together to form the resistances against 13 races (CRY17 to CRY29) and 20 physiological types of wheat stripe rust pathogen *Puccinia striiformis* Westend. f. sp. *Tritici*. A strong spring type was selected to reduce the first developmental stage of the cultivars, so that sowing can be postponed from "Frost's Descent" (late September) to "Winter Beginning" (early November) with much reduced risk of getting stripe rust infection before winter. These two measurements enabled the stripe rust-resistant, high-yield wheat cultivar Fan 6 and its sister lines to remain resistant for 20 years, a world record. Genes for high-quality flour were introduced from abroad to improve the flour quality in the Sichuan wheat production region, and were bred into line 70-5858, which was released in 1970. This elite line is the germplasm for a high flour quality that fits to the Sichuan wheat production region. A "70-5858 × Fan 6" cross was then made to improve the flour quality in the Sichuan wheat production region. Offspring of this cross were given to our sister institute Mianyang Agricultural Research Institute to increase the selection population and thus improve the chance of successful selection. Wheat cultivar Mianyang 11 was bred there, which fulfilled our goal of improving the flour quality in the Sichuan wheat production region. To overcome the shortcoming of lacking resistance to pre-harvest sprouting in Mianyang 11, a sprouting-resistant new synthetic common wheat was made by crossing *Triticum taushii* (Cosson) Schmalh. var. *typicum* (Zhuk.) Yen et J. L. Yang with *Triticum turgidum* L. cv. Ailanmai, a landrace from Jianyang that fits to the eco-environment of Sichuan. Rye (*Secale cereal* L.) germplasm that has multi-spikelets per spike but less grains per spikelet was introduced into common wheat, resulting in the common wheat line "10-A," which has up to 37 spikelets per spike and less grains per spikelet than the ordinary common wheat in the region. All these laid the foundation for further improving wheat in Sichuan.

Another task the author finished during this time was the near completion of the research on the mechanisms in organ constitution of higher plants, the validation of the results, and the writing of the first draft of the manuscript. All these investigations, observations, and thinkings were basically conducted in a "Cowshed" (a kind of informal jail) between events of political criticism and denouncements or reformation through forced labor the authors received there. Here, the author sincerely thanks Ms. Yuzhu Jiang for secretly lending him a microscope disregarding the risks of her own, his lovely wife Prof. Junliang Yang, his son and daughters for secretly providing him the needed books, and papers, pens, and inks for his work, his kindhearted guards for winking at what he was doing. The author could not accomplish this task without these helps. The results of this important research were briefly reported in the *Journal of Sichuan Agricultural University*, 10 (4), 537–565, 1992, under the title "The essential nature of organs in Gramineae, multiple secondary axes theory—A new concept." However, multiple secondary axes as the common basic organ structure of all the higher plants were not reported there. Also, many data and evidence from the observations and investigations and the complete explanation of them were omitted in that brief introduction. Therefore, the author has taken the time to complete this comprehensive writing after he finished his five volumes of *The Biosystematics of Triticeae* in the USA. The author is 93 years old now and does not have much time left in his life and is trying to improve the draft he wrote in the "Cowshed" in Sichuan Agricultural University during the "Great Cultural Revolution" and publish it to contribute to the scientific knowledge of mankind and to write a better period to his life.

Revised in 2014 at Davis, California, USA to avoid interferences Chi Yen
Finalized in 2016 at Yanan, China

Contents

Chapter 1
The Two Mutually Antagonistic Theories About Organogenesis in the Morphology of Higher Plants

Morphology is a basic science of botany. Before the eighteenth century, it mainly describes the morphological characteristics of plant organs. Morphology served taxonomy as a tool for surveying and organizing plant resources. These surveys have continuously enriched the records and descriptions of the morphological types and characteristics of various plant organs.

For morphology itself, the accumulation of such descriptions and the records of various organ morphologies have inevitably led to comparative studies on the morphological structures of each organ to determine their commonness and the differences in their organogenesis and thus their origin and special individualities (i.e., variations). In other words, it tries to answer how the morphology of plant organs has evolved.

In 1790, Goethe published his book "Versuch die Metamorphose der Pflanzen zu erklaren (Studies of the metamorphosis of the plants)," which symbolizes the start of plant morphology as a new research field. His study was based mainly on morphological data. He crafted the term "morphology" for the first time in botanical research. According to Goethe's theoretic analysis, the organs of higher plants are unified. They are all composed of a basic structure: the leaf. Scales, calyxes, petals, stamens, and carpels are all different metamorphic leaves. Leaves extend downward, becoming petioles; petioles merge to form branches. Flowers are metamorphic branches. A plant body consists of multiple units or segments of leaves and petioles. These units merge to form a stem; the stem extends downward, becoming roots. Each of the "leaf and its downwardly extended petiole" structural units was called "phyton" by Gaudichaud-Beaupré (1841) or "anaphyt" by Schultz (1843). This theory is called "phytonic theory" or "anaphyton theory," a concept stating that each higher plant body is built up by interconnecting such leaf organs one after another. This theory was also supported by anatomic studies of the vascular bundle system and phyllotaxy, and it regards stems and roots as derived structures of merged leaf bases, not as the basic organs (Celakovský 1901; Velenovsky 1905, 1913). Hofmeister's hypothesis, as described in his book (Hofmeister 1851) (which

C. Yen, *Constitution of Organs of the Higher Plants*,
https://doi.org/10.1007/978-981-19-1685-4_1

took *Equisetum* as an example and applied it to all higher plants), also falls into this school.

Since the beginning of the nineteenth century, some botanists have believed that axial organs are the basic unit of a plant body and that leaves, which are composed of merged axial branches and webbed structures, are a derivative. This concept was based on palaeobotanical studies of fern (*Polypodiopsida*) and hornwort (*Anthocerotae*) fossils and is antagonistic toward the phytonic theory.

Krüger (1841) suggested that leaves often appear below a growing point and thus are lateral secondary structures and that axial organs are plants' only basic organs. Bower (1884) proposed the concept that the stem is the body axis. Lignier (1903, 1908) hypothesized that original ferns were dichotomously branched and leafless and suggested that because the development of the dominant branch at one side has made it into a sympodial branch, the weak branch changes to a leaf. Therefore, the basic organ is the axial organ, and leaves were transformed from the axes. Bower (1908) proposed his "enation theory," which says that the leaves of *Lycopodiophyta* started from the formation of a thorn-like enation at the side of the stem, which then evolved into flat leaves; the vascular bundles differentiated from the stele reach the bases of the flattened leaf and finally evolve into a single vein structure, extending to the leaf apex. Lang (1931) and Bower (1935) further stated that *Psilophyton* represents the thorn-like enation stage during the leaf evolution, *Thurstophyton* represents the flat-leaf stage, *Asteroxylon* represents the stage when the vascular bundles reach the base of the flat leaf, and *Arthrostigma* represents the stage when the vascular tube extends to the leaf apex.

The concept of axial origin of higher plant organs was adapted to explain *Psilophyton's* origin in the telome theory of Zimmermann (1930, 1938). Based on the fossil records of *Psilophyton*, he termed the *Rhynia*-like upright branch structure "telome" (terminal branch) and the branch below telome "mesome." Telome and mesome compose the "telome truss" (Fig. 1.1a). A telome can be sterile or fertile if a sporangium developed on the apex. The telome theory states that dichotomous telome trusses become a sympodial branching structure through overtopping (Fig. 1.1b); the dominant branch develops into a stem, while the weak branch stops growing and becomes a dense telome truss through reduction and aggregation (Fig. 1.1c); and then the dense telome truss arranges into a plane with branches webbed together and having parenchyma. This flat, dense telome truss becomes a macrophyll by webbing (Fig. 1.1d), and the aggregated telome truss forms the veins. Therefore, the telome theory believes that primitive veins should be dichotomously open-ended and should be a direct extension of the stem, an equivalent to the stem.

The telome theory believes that *Lycopodiophyta*'s microphyll is also formed from a single flattened telome truss, which is different from Brower's enation theory.

According to the telome theory, macrophylls are formed through planation, aggregation, flattening, and webbing of the weak sympodial branches, and their most primitive phyllotaxy should be alternative. It is aggregation that has made them opposite. Axillary buds are also formed by lowering the position of telome aggregation to the axillaries. Sporangia form at the apexes of telomes. The most primitive sporangium forms at the apex of the marginal vein of the macrophyll during leaf

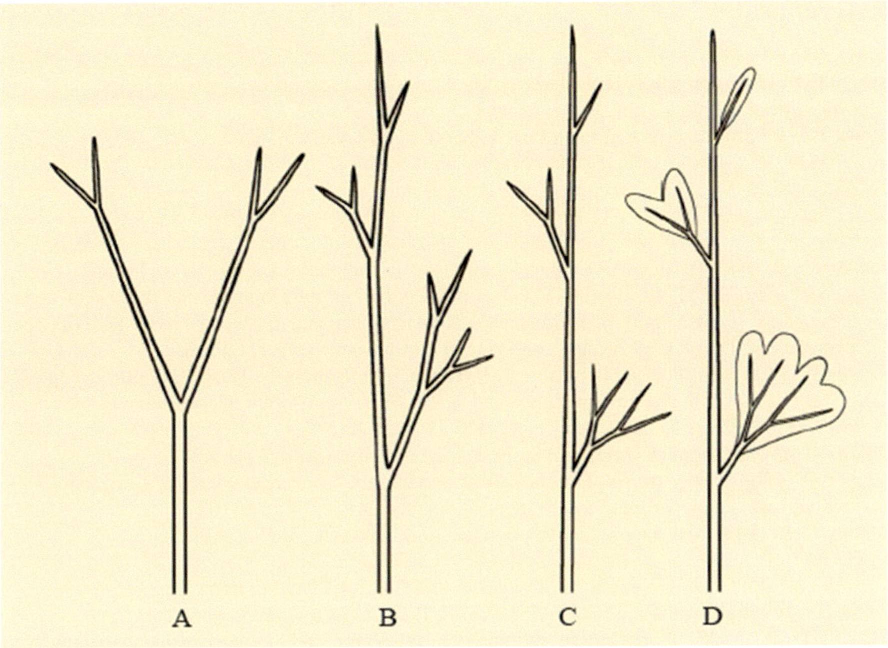

Fig. 1.1 The evolution of macrophyll according to the telome theory. (**a**) A telome truss consisting of telome and mesome, (**b**) sympodia formed from weak and dominant dichotomous branches via overtopping, (**c**) aggregation by shortening lateral branches and the mesomes, and (**d**) leaves formed by webbing the flattened lateral branches

evolution. In the case of *Lycopodiophyta's* microphyll, however, they are located anywhere on the leaf surface where single fertile telomes aggregate.

According to the explanation by the telome theory, the sporophyll or the sporangiophore of *Equisetum* form by merging the tops of the outward-turning dichotomous branches, and bracteoles are evolved from sterile telomes (Scott 1900).

Smith (1955), who advocated the telome theory, could not answer how *Equisetum*'s cyclic phyllotaxy has evolved. It cannot be explained by the telome theory, and his only explanation was "do not know yet." *Sphenophyllum* and its relatives have cuneate leaves with dichotomous veins but no petioles. This kind of leaves seem to be cladode. You can see single-veined sterile leaves on many *Equisetum* species. They may be a form of degenerated sterile cladode.

In terms of the morphology of the organs of higher plants, the aforementioned are at present the two main antagonistic viewpoints about the units of organ structure and their basic morphological composition. Although small arguments happen here and there among researchers, in general, there is nothing but the phytonic theory, which regards leaf organs as the basic structural unit for segmental composition, and the antagonistic telome theory, which believes in the continued development of axial organs from the base to the top without segmentation. Currently, the telome theory is adopted by the majority and is the dominant theory in plant morphology.

References

Bower FO (1884) On the comparative morphology of the leaf in the vascular cryptogams and gymnosperms. Philos Trans R Soc Lond 175(Part 2):565–615
Bower FO (1908) The origin of land flora. Macmillan, London
Bower FO (1935) Primitive land plant. Macmillan, London
Celakovský LJ (1901) The structure of Kaulome. Botanische Zeitung 59(1):79–144
Gaudichaud-Beaupré MC (1841) Voy. Bonite, Bot, vol 3. Memories of the Academy of Sciences, Paris
von Goethe JW (1790) Studies of the metamorphosis of the plants. Carl Wilhelm Ettinger, Gotha
Hofmeister WFB (1851) Comparative studies of germination, development and fruiting of higher cryptogams (moose. Ferren, Euisetaceen, Rhizocorpeen and Lycopcdiaceen) and the seed formation of the conifers. Verlag von Fridrich Hofmeister. Kruger, Leipzig
Krüger MS (1841) Bibliographia botanica[M]. Heude und Spenersche Buchhandlung, Berlin
Lang WH (1931) On the spine's sporangia and spores of *Psilophyton* princeps shown in specimens from Gaspe. Philos Trans R Soc Lond Ser B 219:421–442
Lignier O (1903) Origin of Pteridophyta. Bull Soc Linn Normandie 5 ser 7:93
Lignier O (1908) Essay on the morphological evolution of the vegetables. In: Congress of the French Association for the Advancement of Science, Reviews, pp 53–62
Schultz CH (1843) The anaphytosis or rejuvenation of plants. Verlag von August Hirschwald, Berlin
Scott DH (1900) Studies in fossil botany. Adam and Charles Black, London
Smith GM (1955) Cryptogamic botany[M], vol II, II edn. McGraw-Hill, New York
Velenovsky J (1905) Comparative morphology of plants, vol I. Verlagsbuchhandlung von Fr. Rivnáé, Prag
Velenovsky J (1913) Comparative morphology of plants, vol 4. Verlagsbuchhandlung von Fr. Rivnáé, Prag
Zimmermann W (1930) The phylogeny of the plants. Verlag von Gustav Fischer, Jena
Zimmermann W (1938) The telome theory. Der Biologe 7(12):385–390

Chapter 2
Saunders' Leaf-Skin Theory of the Stem

In 1922, Edith R. Saunders of the University of Cambridge published an article entitled "The Leaf-Skin Theory of the Stem: A Consideration of Certain Anatomico-Physiological Relations in the Spermophyte Shoot" in Volume 36 of Annals of Botany. The leaf-skin theory of the stem that he proposed in this article first appeared in his mind in his study of the hairs on the seedlings of *Matthiola incana* and was later concluded after his observations of other spermophytes. This theory states that the surface tissue of the shoot axis of all spermophytes is formed by downward growth and the extension of the leaf petioles, which then fuse around the shoot axis. He believed that this appears to be the case also in some of the Pteridophyta at least. This theory is highlighted here because it is based on objective experimental evidence.

Saunders noticed that the hypocotyl of a *Matthiola incana* seedling was hairless, which differed from the plumule but had the same morphology as the plumule. He further observed similar phenomena on the seedlings of other plants, such as *Veronica hederacfolia*, which has a line of hairs along the edges of the cotyledon extensions of a hypocotyl, but the hypocotyl is morphologically similar to internodes and *Ipomoea sanguinea*, whose cotyledons extend downward and form a furrow on the hypocotyl (Fig. 2.1). He believed that the cotyledons extend downward around the shoot axis, forming the surface of the hypocotyl. The stem above the cotyledons is also formed by the extension of the petiole, which wraps the shoot axis (Figs. 2.2 and 2.3).

Based on these observations and analyses, Saunders made 19 conclusions. The following are the summary and explanations: the surface tissue of the shoot axes of spermophytes are of foliar origin; i.e., the downward extension of leaves is not only observed in those obvious cases but also generally appears in all spermophytes and, at least, in some of the Pteridophyta. In these species, the surface tissues of the hypocotyl are extensions from the cotyledon because they have the same characteristics on their surface. Foliar skin is formed through a downward growth and the extension of the leaf primordium wrapping the shoot axis and forming the surface of the pith. The contact edges of the extensions may be perfectly aligned or deeply

C. Yen, *Constitution of Organs of the Higher Plants*,
https://doi.org/10.1007/978-981-19-1685-4_2

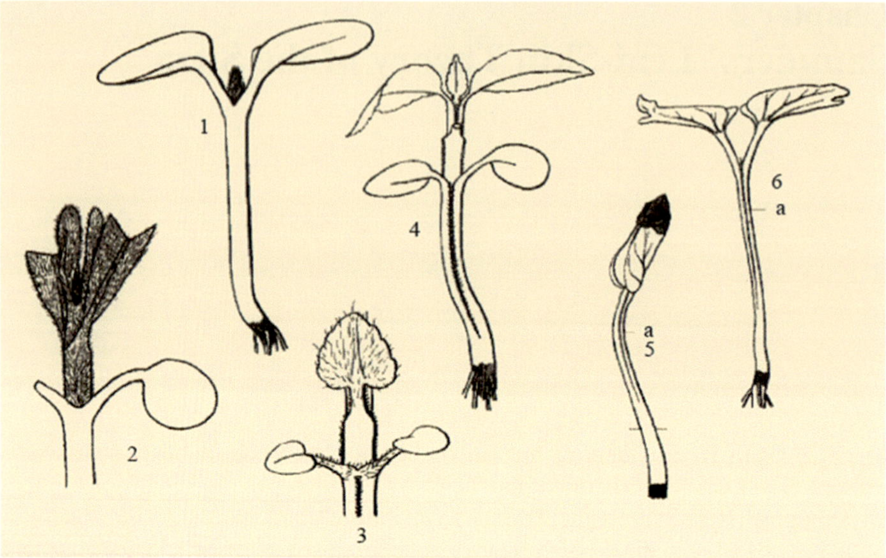

Fig. 2.1 Dicotyledonous seedlings, showing the leaf-skin relationship of cotyledons, leaves, hypocotyls, and internodes. (1) *Matthiola incana* seedling, showing glabrous cotyledon and hypocotyl and hairy plumule; (2) older stage with the removal of the lower portion of hypocotyl, the first cotyledon, and the upper portion of the first pair of leaves; (3) *Veronica hederacfolia* seedling, showing hypocotyl with a line of hairs along the cotyledon extensions; (4) *Lopezia coronata* seedling, showing the same feature on hypocotyl and internodes; (5) *Ipomoea sanguinea*, showing young seedling with the furrow (a) formed between the edges of the cotyledon extensions; (6) *Ipomoea sanguinea* seedling at older stage with open cotyledons. (After Figures 1–6 in Saunders 1922)

fused with each other without any visible trace, or they may exhibit a characteristic edge with a boarder in different forms (e.g., ridges, furrows, lines of hairs, or different colors). When a leaf-insertion completely encircles the stem (as in most monocotyledons and a few dicotyledons) or when opposite or whorled leaf insertions come into lateral contact (e.g., many dicotyledons and some conifers), the downward extension is limited to one internode; when the insertion width of the opposite leaves is less than ½ of the circumference of the stem (especially some dicotyledons with four-angled stems), the extension stretches through two internodes; when the insertion of spirally arranged leaves is not wider than the circumference of the stem, the number of internodes that the extension can go through depends upon the distance between the leaves and the leaf insertion width (Fig. 2.4).

All in all, Saunders believed that the leaf is an independent organ, and the skin of the stem is the downward extension of the leaves. Saunders' theory basically follows the phytonic theory. The difference between them is that, he believed, there is a continuously growing inner stem axis and that the downward extension of the leaves is only limited to the skin of the stem.

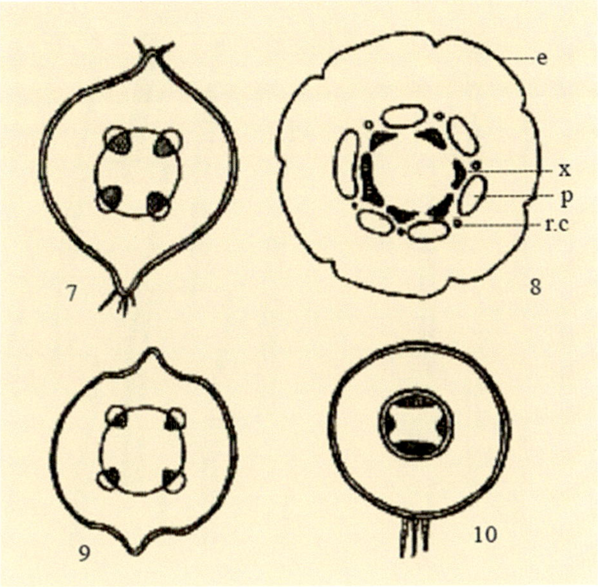

Fig. 2.2 Transverse sections of the stem of a young dicotyledonous seedling, showing the relationship between the leaf-skin and the central axis. (7) A transverse section of *Ruellia amoena* hypocotyl, showing the potential edges of the cotyledon extensions that are thrown up into a ridge surmounted by hairs; (8) a transverse section of *Pinus maritima* hypocotyl (e: epidermis, x: xylem, p: phloem, r and c: resin canal); (9) a transverse section of *Rivina humilis* hypocotyl, showing the same feature as *Rucllia amocna* but without the hairs; (10) a transverse section of *Stellaria media*. (After Figures 7–10 in Saunders 1922)

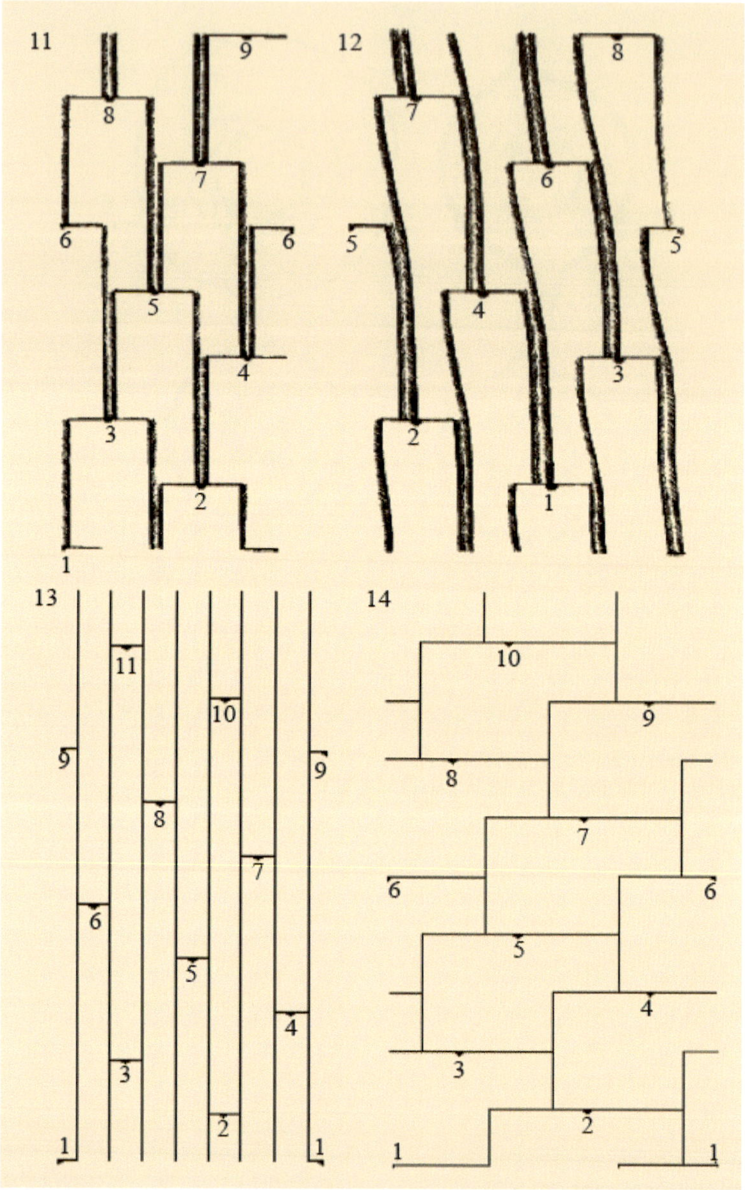

Fig. 2.3 An illustration of the phyllotaxis of dicotyledons, showing the axial view as if it is longitudinally split and spread out flatly. (11) *Veronica hederacfolia*, leaf divergence 2/5, leaf insertion approximately 2/5; (12) *Lopezia coronata*, leaf divergence 2/5, leaf insertion approximately 1/5; the pattern shows a lateral skew; (13) *Reseda odorata*, flowering axis; (14) bract divergence 3/5, bract insertion approximately 1/2. (After Figures 11–14 in Saunders 1922)

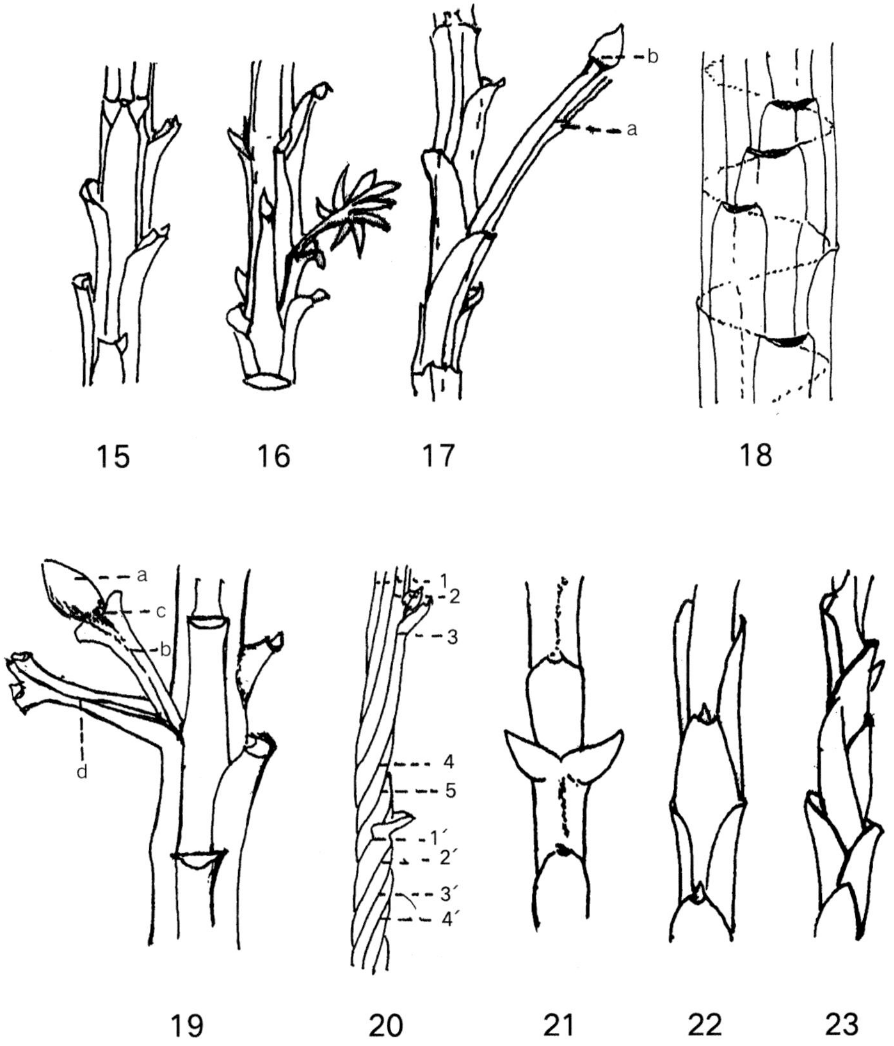

Fig. 2.4 An illustration of dicotyledonous phyllotaxis. (15) *Reseda odorata*, a vegetative axis; leaf divergence 2/5, leaf insertion approximately between 1/5 and 2/5; (16) *Reseda odorata*, flowering axis; leaf divergence 3/5, close to 2/5; the pedicel showing the sepal extensions; (17) *Viola tricolor*, showing contours of bract and sepal extensions, (a) a bract and (b) a posterior sepal (the appendiculate portion has been cut away); (18–20) *Calystegia dahurica* (18: stem viewed as if transparent; the phyllotaxis is the same as in *Reseda odorata* (15); 19: (a) an unopened flower bud, (b) contour lines from the bracts that have been cut away, and (c, d) a channeled petiole; 20: torsional leaf skin); (21) *Epilobium parviflorum*, glabrous cotyledons, hypocotyl ridge without hairs, ciliated petioles, internodes with ridge line surmounted with hairs; (22 and 23) *Lopezia coronate*, showing leaf extension with a line of hairs on edges (22: lower region of the stem with opposite leaves, 23: upper region with spirally arranged leaves)

Reference

Saunders ER (1922) The leaf-skin theory of the stem: a consideration of certain anatomico-physiological relations in the spermophyte shoot. Ann Bot 36:135–165

Chapter 3
Leaves Are Split Axes

In common wheat (*Triticum aestivum* L.), an extra spikelet often develops at the spike node below a spikelet (Figs. 3.1 and 3.2). This kind of spikelet is called a "supernumerary spikelet." The importance of this is that such a spikelet may undergo various evolutionary stages and thus vary in morphology. A supernumerary spikelet starts as a small protuberance and then develops into a small spine; the spine can further develop into a thick awn (axis) and then into a flat, tongue-like structure with enhanced photosynthetic activity, a function of a leaf; the tongue-like leaf may continue to enlarge its surface area and become a capsular structure, which may further enlarge and then split to become a flat glume (equivalent to a green leaf). Here we see a very clear evolutionary process: the formation of a phyllome from an axial structure through enlargement and slitting. The second growing point occurs at the bottom inside the capsule, where the second glume and then a supernumerary spikelet (an equivalent to a branch) develop following the same evolutionary process. This series of evolutionary stages represented by these factual cases is not in conformity with either the phytonic theory or the telome theory. It does not agree with the phytonic theory because these cases suggest that an organ is of axial origin. It does not agree with the telome theory since these cases show that a stem is not an extension of the axis but an interconnection of many secondary axes. The fact here is that a leaf is a split axis. This concept completely differs from all the existing theories. Therefore, how the organs of higher plants are formed and constituted must be further studied.

The above examples of the morphological evolution of the supernumerary spikelets in wheat obviously show that phyllomes are formed through enlargement, splitting, and planation of axes and that stems are composed of interconnected secondary axes. Leaves originate from split axes. Although a complex leaf may be composed of multiple split and unsplit secondary axes, axis splitting is their basic feature. This concept of the axial origin of organs is consistent with the telome theory, but this concept of leaf formation is principally different from the telome theory. The concept that a stem is composed of interconnected axial segments seems to have some similarities with the phytonic theory, but it is actually different from

C. Yen, *Constitution of Organs of the Higher Plants*,
https://doi.org/10.1007/978-981-19-1685-4_3

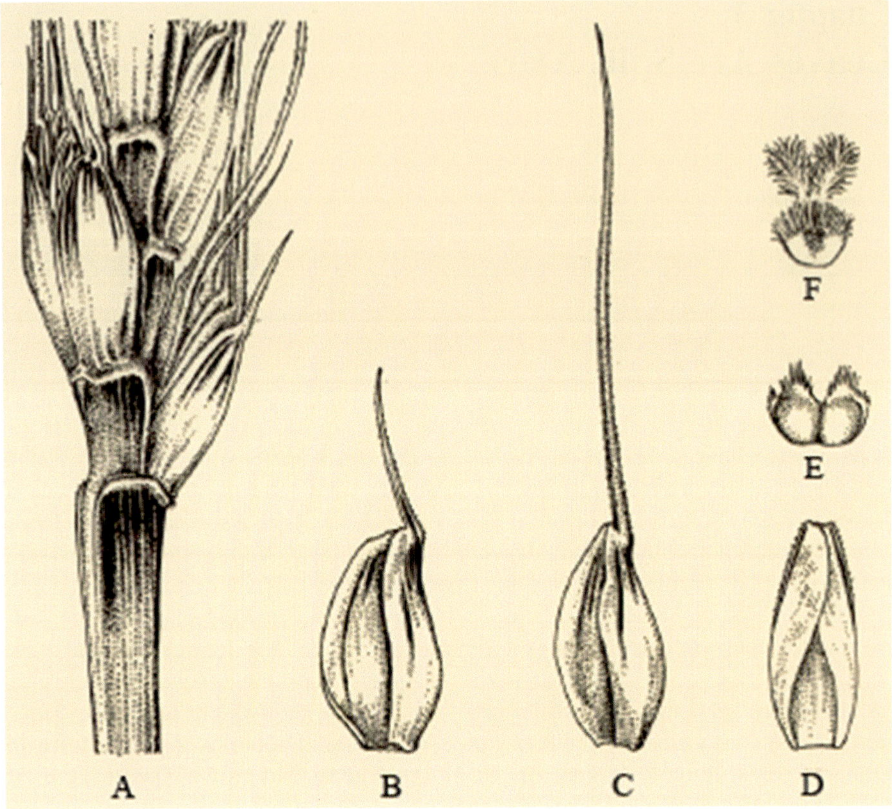

Fig. 3.1 Spike organs of *Triticum aestivum* L. (**a**) Lower portion of the spike, showing the morphological structures of the spike axis and the spikelets (one can clearly see a complex node structure and the upper internode buried into the lower internode); (**b**) glume; (**c**) lemma; (**d**) palea; (**e**) scale; (**f**) pistil

the latter in principle because stems are regarded as one of axial origin here. This concept is also principally different from the telome theory because we regard stems as a composition of secondary axial segments. The organogenesis of wheat supernumerary spikelets is direct evidence of leaf cleft axes. As shown in Fig. 3.1a, a spiked axe (i.e., the stem) is clearly segmental, with the upper internode buried into the lower internode. Figure 3.2 shows the various developmental structures of supernumerary spikelets. There is no doubt that b and c are of typical axial structure; d and e show the further evolution of the axial structure through expansion, enlargement, and planation; f and g show the splitting caused by the enlargement, forming a primitive glume; h and i illustrate the further evolution of a protoglume into a supernumerary spikelet. The phyllomes above the first glume on the spikelet are all reformed at the bottom inside the split axial capsule one after another, following the same model exactly. The inside of the bottom of the split axial capsule is also the place where auxin accumulates. Thus, it is reasonable for the auxin to facilitate the

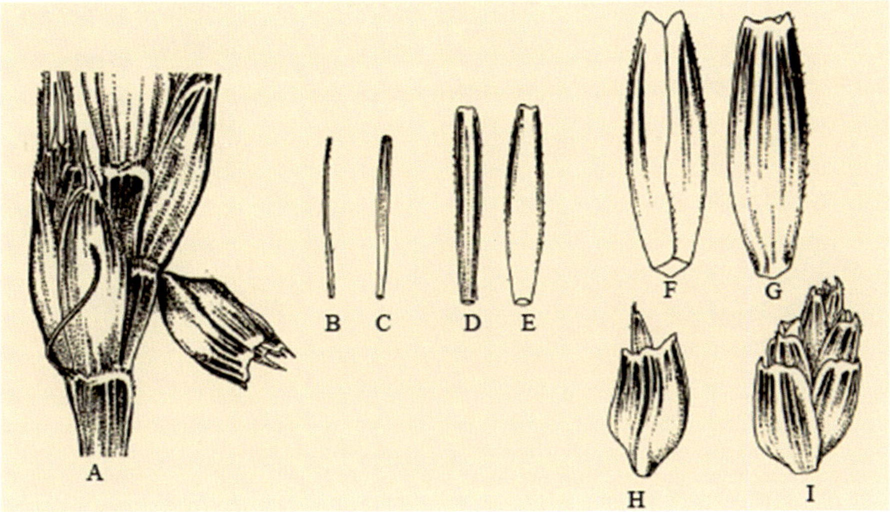

Fig. 3.2 The supernumerary spikelets of *Triticum aestivum* L. (**a**) Part of a spike with a supernumerary spikelet, showing where the supernumerary spikelet grows; (**b**) an awn (axis)-like spine; (**c**) flattened and expanded supernumerary spikelet; (**d**) a further expansion of the flattened supernumerary spikelet, forming a small cavity at the middle; (**e**) a further expansion of the sac-like supernumerary spikelet without splitting; (**f**) ventral view of a splitting sac-like supernumerary spikelet; (**g**) dorsal view of a splitting sac-like supernumerary spikelet; (**h**) a second glume (the second axis) is formed at the inner bottom of the first glume; (**i**) a well-developed supernumerary spikelet (branch)

formation of the secondary growing points there. Figure 3.3 clearly shows the developmental process from an axis to a leaf cleft and from a single vascular bundle to multiple vascular bundles. Vascular bundles are secondary. This is a fact and a direct piece of evidence without a doubt.

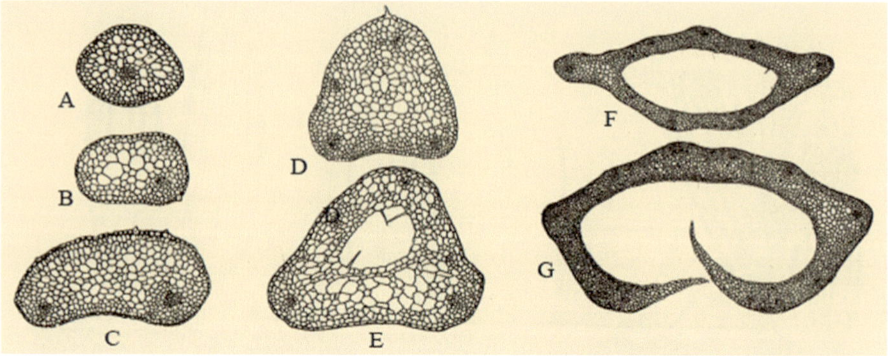

Fig. 3.3 Transections of the supernumerary spikelet of *Triticum aestivum* L. at various developmental stages, showing the evolutionary process from axis to phyllome at various stages. (**a**) Spine stage with a single vascular bundle in the well-developed central portion; (**b**) a spine-like supernumerary spikelet with initial flat expansion; the central vascular bundle has moved toward one ridge, and the development of vascular tissues has been initiated near another ridge; (**c**) a flattened, expanded spine-like supernumerary spikelet with a vascular bundle in each ridge; (**d**) a further expanded flat supernumerary spikelet with a small hollow center; (**e**) a further expanded flat supernumerary spikelet having an expanded hollow center with hairy cells inside; (**f**) a sac-like supernumerary spikelet with a large hollow center; (**g**) the sac-like supernumerary spikelet has split into a glume, finishing the evolution from an axis into a leaf

Chapter 4
Reevaluation of the Evidence for the Telome Theory

The telome theory is based on the fossils of *Psilophyta* and *Protopteridales*. These fossils objectively exist. Their objective existence itself is not contradictory. The contradiction comes from understanding their objective existence. In other words, the contradiction is not about the existence itself but about our subjective understanding of the existence.

Many pieces of evidence show that spermophytes (seed plants) were derived from Pteridophyta (ferns). There is basically no argument about this. The phylogenetic arguments are focused on the origin of Pteridophyta. The telome theory is also about organogenesis during the evolution of Pteridophyta.

There are two different viewpoints about the origin of Pteridophyta: from Phaeophyceae (Church 1919; Arnold 1947) or Chlorophyta (Bohlin 1901; Lotsy 1909; Fritsch 1916) and from Bryophyte. However, there is more evidence showing that Pteridophyta are closer to the mosses. Zimmermann (1930, 1938), who proposed the telome theory, believed that Pteridophyta were phylogenetically close to Bryophyte; both originated from primitive terrestrial Archegoniatae but had since evolved through different routes. Supporters of the telome theory who believe that Psilophyta are the most primitive also believe that Pteridophyta have a close tie with Anthocerotophyta (hornworts), which have alternative generations with a leaf-like gametophyte and sporophyte or embryos that are partially parasitized on the gametophyte.

Campbell (1899) once pointed out that the sporophyte of hornworts has the ability of infinite growth, has highly developed assimilating tissue, and is very like the independent sporophyte of ferns. Hornworts' sporophyte is considerably independent. Its foot can either stretch into the soil like a root with a special assimilating organ or develop into a leaf. Campbell (1924) reported that under very favorable growth conditions, hornworts' sporophyte has a finite spore-forming structure on the apex, while conducting tissues are differentiated at the lower part of the column axis. He believed that hornworts' meristem zone would have been morphologically similar to *Rhynia* if the meristem zone had moved from the base to the apex of the axial structure, forming a finite sporangium structure there while creating

C. Yen, *Constitution of Organs of the Higher Plants*,
https://doi.org/10.1007/978-981-19-1685-4_4

Fig. 4.1 *Rhynia major* Kidston et Lang. (After Figure 96 in Smith 1955, which was adapted from Kidston and Lang 1921)

dichotomous branches via bifid division. However, this hypothetic transition is based entirely on speculation without any supportive specimen.

Smith (1955) pointed out many similarities of the sexual organs between ferns and hornworts. For instance, both have their sexual organs embedded in the gametophyte without archegoniophore. Also, hornworts' archegonia and antheridia are structurally similar to those of modern ferns, such as *Psilotum* and *Tmesipteris* (Fig. 4.1). Therefore, *Psilophyta* is believed to be the most primitive fern associated with bryophytes.

Another important argument for the telome theory is the ancient nature of Pteridophyta. The most ancient ferns uncovered in the strata are Pteridophyta, which can be traced to the Silurian. *Rhynia*, which grew ferociously in the Devonian period and had dichotomous branches but no leaves, no roots, and only rhizomes and aerial stems with apical sporangia, is believed to be the most primitive model structure (Fig. 4.1). However, these so-believed most primitive ferns are actually not the most ancient ferns. The most ancient Pteridophyta species is the leafy *Zosterophyllum*, which were extensively distributed throughout the strata of middle Silurian from Scotland to Australia and had inverted and spiked sporangia at the branch apex. Also found in the middle Silurian strata in Australia is the fossil of *Baragwanathia longifolia*, another leafy lycophyte that belongs to *Protolepidodengraceae*. Another Protolepidodengraceae species,

Protolepidodendron, lived in the same period as *Rhynia's*, that is, in the middle or late Devonian, but has a much wider distribution area than *Rhynia*. *Rhynia* has only been found in the Devonian strata in Scotland. *Protolepidodendron* was found not only in Scotland but also in France, Czech's Bohemia, and the Eastern United States of America. These indicate that Pteridophyta, particularly *Rhynia*, are not the most ancient ferns. Leafy *Baragwanathia* is at least a plant species living in the same period as the most primitive Pteridophyta species, *Zosterophyllum*. Rather than saying that the ancient leafy *Baragwanathia* evolved from the later-emerged leafless *Rhynia*, it is better to say that the leafless *Rhynia* was derived from the earlier-emerged leafy ferns via leaf degeneration.

Pteridophyta are said to have evolved from the common ancestor of hornworts. However, whether they were leafy or leafless ancient ferns, their morphological structures are still very different from hornworts. The ancient ferns are structurally far more complex than hornworts. The fossils of leafy ferns and the later-emerged leafless ferns coexisted in the Middle Silurian strata. This indicates that there still are many undiscovered evolutionary steps between them and their distant ancestor and that neither of them is the most ancient. The fossil record is incomplete, and there must be a lost link between these ancient ferns and hornworts that is yet to be discovered.

From the perspective of the modern Psilophyta species, both *Psilotum* and *Tmespteris* are leafy. *Psilotum* leaves are very small, just like scales, while the leaves of *Tmesipteris* are very large. This indicates that the majority of Psilophyta species have leaves (Fig. 4.2), although the leaves are so-called "microphylls," not "macrophylls," as described under the telome theory; they are unlikely constructed through the morphogenesis of aggregation, flattening, and webbing of the apical branches, as stated by the telome theory. The mode of aggregation, flattening, and webbing of the apical branch is assumed and is not supported by objective evidence. Leaves should not be divided into the so-called "microphylls" and "macrophylls" because they only differ in size and complexity.

Comparing the structure of Psilophyta with the fossils of Protolepidodengraceae, it is reasonable to regard the leafless Psilophyta taxa as a manifestation of degradation from the leafy taxa.

The seedlings of *Tmesipteris tanneusis* have two leaves or three verticillate leaves at the base of their aerophilic branches with apical growing point at the middle of leaf axillae. Their upper leaves become alternate phyllotaxy due to unbalanced development. Both the triplet short sporangium branch of *Psilotum triquetrum* and the sporangium of *Tmesipteris tanneusis* are axillary. This indicates that branching is of axilla but not of dichotomy, which forms by the division of the growing point, as described in the telome theory. This is consistent with the morphological construction mode of multiple secondary axes.

The "macrophyll" of *Protopteridium minutum* and *Cladoxylon seceparium* can be regarded as a simple (monophyllous) leaf formed through divisions and flattening of multiple secondary axes, just like the branching "microphyll" of *Protolepidodendron scharyanum*. According to the multiple secondary axis theory,

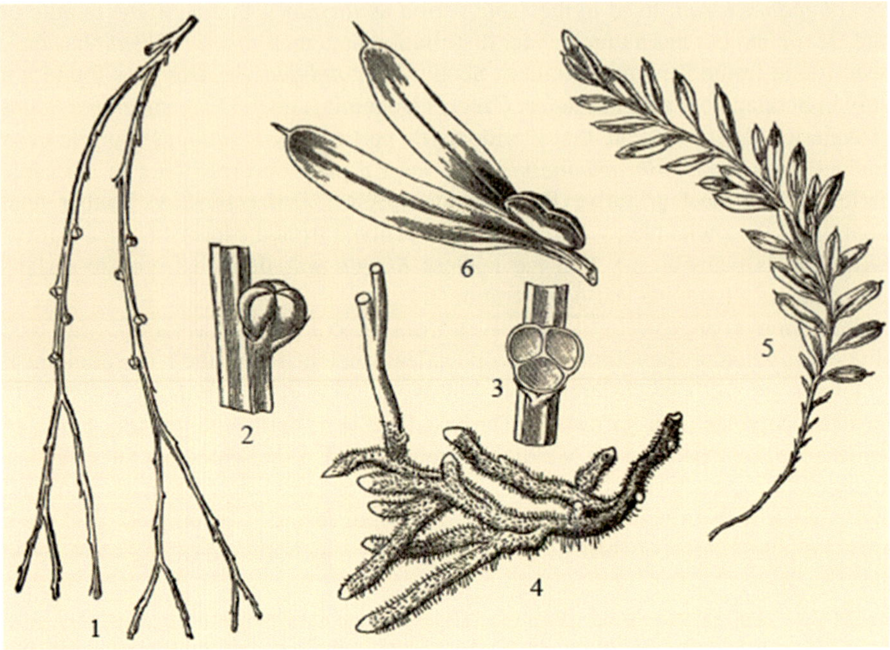

Fig. 4.2 *Psilotum nudum* and *Tmesipteris tanneusis*. (1) Overhanging branches of *Psilotum nudum*; (2) a sporangium of *Psilotum nudum* grown at two-lobed leaf axilla; (3) the transection of the sporangium illustrated in image 2 showing three chambers inside the capsule; (4) a rhizome of *Psilotum nudum*; (5) a leafy branch of *Tmesipteris tanneusis*; (6) a sporangium of *Tmesipteris tanneusis* grown at two-lobed leaf axilla. (After Figure 316 in Ikeno 1906)

Pteridophyta's first leaf is the monophyllous leaf commonly called cotyledon. This view is also irrefutable.

The sporangia of the *Protopteridium minutum* are at the leaf's apex, which is also not a problem, in principle, from the perspective of the multiple secondary axis theory. The locations of sporangia should be decided mainly by physiological factors and are determined physiologically and genetically in different species. The sporangia of *Protolepidodendron scharyanum* are located at the surface of fertile leaves, where the central vascular bundle passes by, and not at the leaf apex nor at the tip of the veins. The belief that the branch-tip location of sporangia is the most primitive morphological structure is the subjective viewpoint of the telome theory, which considered *Rhynia major* Kidston et Lang as the most primitive model of ferns and speculated the aggregation and webbing of the telome.

The leafless ferns can be regarded as an evolutionary side branch of the leafy ferns due to the degradation of the leaf clefts. This type of degradation is not uncommon in higher plants. The telome theory cannot explain the axially splitting nature of the leaves and the axillary origin of the secondary axes (formed at the bottom of the hollow center) during the morphogenesis of supernumerary spikelets of wheat. It is unable to explain the multilevel joint structure of the stem either. However, the multiple secondary axis theory is able to explain these by pointing out that the morphological constitution of leafless ferns results from the degradation of leaf clefts. The traces of leaves disappear as the leaves degrade, and the interconnected secondary axes become morphologically continuous without an overall segmental appearance. We do not have fossil data about the development of gametophyte seedlings of ancient Psilophyta species. The gametophytes of the modern Psilophyta species, such as *Psilotum* and *Tmesipteris*, do not have chlorophyll in the first place; they, just like a degenerated mycorrhiza, live underground saprophytically or symbiotically with algae. Therefore, they were not the most primitive ferns either.

According to the subjective inference of the telome theory, the "macrophyll" formed through overtopping are alternate leaves and thus the most primitive, while opposite leaves are secondary, forming through the aggregation of the alternate. This inference, too, contradicts all observed instances of ontogenesis. During the ontogenesis of all higher plants, except those with single-lobed leaves, opposite phyllotaxy is formed via an unbalanced development of alternate or cyclic phyllotaxy by spiraling one leaf up or degenerating another leaf down. All higher plants, including the modern Psilophyta species *Psilotum* and *Tmesipteris*, are so developed. The telome theory cannot explain these facts.

According to the speculation under the telome theory, the central axis of a vascular bundle is original. The lateral branches aggregate into leaf veins after the telome dichotomously overtops. The leaf veins then web to form a leaf. Leaf veins precede the formation of the leaf, which is contrary to the observed leaf development in all higher plants. During the ontogenesis of all higher plants, leaf buds do not have any vascular bundle but have parenchyma cells at their early developmental stage. Vascular bundles begin to form at the base of a leaf primordium, and the bipolar differentiation continues gradually toward the apex to form the leaf veins and downward to form the vascular bundles in the base, which join those of the next segment to form the central axis of the stem.

As early as 1949, experiments by G. Gamus showed that the buds grafted on a fleshy carrot root induced the differentiation of a layer of parenchyma tissue into a procambium. Treating a fleshly cut carrot root with synthetic auxin can also induce the parenchyma tissue to form a procambium. These experiments clearly indicate that the differentiation between a procambium and parenchyma tissue is due to the function of auxin in the presence of sugar. Auxin at proper concentration can induce the formation of the procambium from parenchyma tissue, which then develops into a vascular bundle. The auxin produced in a leaf bud is transported downward and accumulates at the base of the leaf bud and induces the formation of the vascular bundle there. Leaf veins are differentiated tissue induced by the flux of a high concentration of auxin along its way. Therefore, it is easy to understand why leaf veins can be parallel or merged to become dichotomous or meshed.

From the conclusions of modern plant physiological experiments (Wetmore and Soeokin 1955; Torrey and Thimann 1972), the differentiation between vascular bundles is the action of auxin and sucrose. Their presence and concentration determine the quality and quantity of the differentiation. Meanwhile, experiments have also shown that cytokinins and gibberellins play a decisive role in xylem differentiation. Vascular bundles and xylem are both secondary structures evolved from the primary parenchyma under the induction of certain chemical compounds. These specialized secondary transportation issues connect multiple secondary axes into a whole, making them unite into a compound body with uniformed physiological regulation. This compound body is called "stem." For any higher plant, whether it is grass, a shrub, or a tree, the stems are this kind of secondary structure.

References

Arnold CA (1947) Introduction to paleobotany. McGraw-Hill, New York
Bohlin K (1901) Draft of the phylogeny of the green algae and archegonia. Almquist & Wiksell, Uppsala
Campbell DH (1899) Lectures on the evolution of plants. Macmillan, New York
Campbell DH (1924) Self-sustaining sporophytes of Bryophyta. Ann Bot 38:473–482
Church AH (1919) Thalassiophyta and the subaerial transmigration. Oxford University Press, London
Fritsch FE (1916) The algal ancestry of the higher plants. New Phytol 15:233–250
Gamus G (1949) Studies on the role of buds in morphogenetic phenomena. Rev Cytol Paris 11:1–199
Ikeno S (1906) Plant systematics (Chinese translation by Luo Zongluo, published in 1947 published by The Commercial Press, Shanghai)
Kidston R, Lang WH (1921) On old red sandstone plants showing structure, from the Rhynie Chert Bed, Aberdeenshire. Part IV. Restorations of the vascular cryptogams, and discussion of their bearing on the general morphology of the Pteridophyta and the origin of the organisation of land plants. Trans R Soc Edinburgh 52:831–854
Lotsy JP (1909) Lectures on botanical tribal history, vol 2. Verlag Gustav Fischer, Jena
Smith GM (1955) Cryptogamic botany, vol II, II edn. McGraw-Hill, New York
Torrey G, Thimann KV (1972) On the beginnings of modern studies of plant morphogenesis: a tribute to Ralph G. Wetmore's contributions. Bot Gaz 133(3):199–206
Wetmore RH, Soeokin S (1955) On the differentiation of xylem. J Arnold Arboretum 36:305–317
Zimmermann W (1930) The phylogeny of the plants. Verlag von Gustav Fischer, Jena
Zimmermann W (1938) The telome theory. Der Biologe 7(12):385–390

Chapter 5
What Do Hooded-Awn Barley, Bitter Bamboo, Tortoise Shell Bamboo, Paspalum, and Crabgrass Show Us?

Barley (*Hordeum vulgare* L.) has some special varieties, such as those commonly called hooded-awn barley, which have abnormal awns. Figure 5.1 shows the developmental stages of the leaf of hooded-awn barley, which demonstrates that a leaf of Gramineae is a compound leaf, a kind of macrophyll that forms via deformation and convergence of multiple secondary axes grown on one slit axis. In this regard, a comparative study of these structures with those of bamboo leaves, which have well-defined inner and outer ligules, will clearly demonstrate their evolution (Figs. 5.2 and 5.3).

As shown in Fig. 5.1g, barley's hooded awn is a branch, which, just like a wheat spike with supernumerary spikelets, has a compound node, on which two back-split supernumerary spikelets grow in an axillary manner. This phenomenon well demonstrates that the awn on a glume (as a splitting axis or a leaf) is a very complex branch that has evolved from multiple secondary axes. The apparently simple constriction usually seen on a leaf of Gramineae is actually a compound node, like the compound node on a wheat spike, that has evolved from multiple secondary axes and from which supernumerary spikelets can develop. Such a compound node can further degenerate to a translucent node of a hooded awn. When the awn (i.e., converged secondary axes) then flattens to become a macrophyll, the awn node degenerates into a constriction. From the specimens shown in Fig. 5.1g–i, we can clearly see that the two back-split protoglumes of supernumerary spikelets converge at the back and then split again to the left and right to form two half-split axes, which then converge to construct two ligules. From these real images, we can see that the structure of a Gramineous macrophyll, such as those of barley, wheat, and bamboo, is very complex. They are made not by a simple axial split but through a series of evolutionary processes in which multiple secondary axes grow, converge, split, degenerate, expand, and regrow. For a leaf, whether the upper secondary axes split or not, the lowest secondary axis, as the base, will always split. Monocotyledons undergo a single split; all their secondary axes, but ratoons and callus, originate from the axis, which splits at the inner side, i.e., in an axillary manner. The constrictions of barley hooded awns and leaves, wheat leaves, bamboo leaves, and their spike and

C. Yen, *Constitution of Organs of the Higher Plants*,
https://doi.org/10.1007/978-981-19-1685-4_5

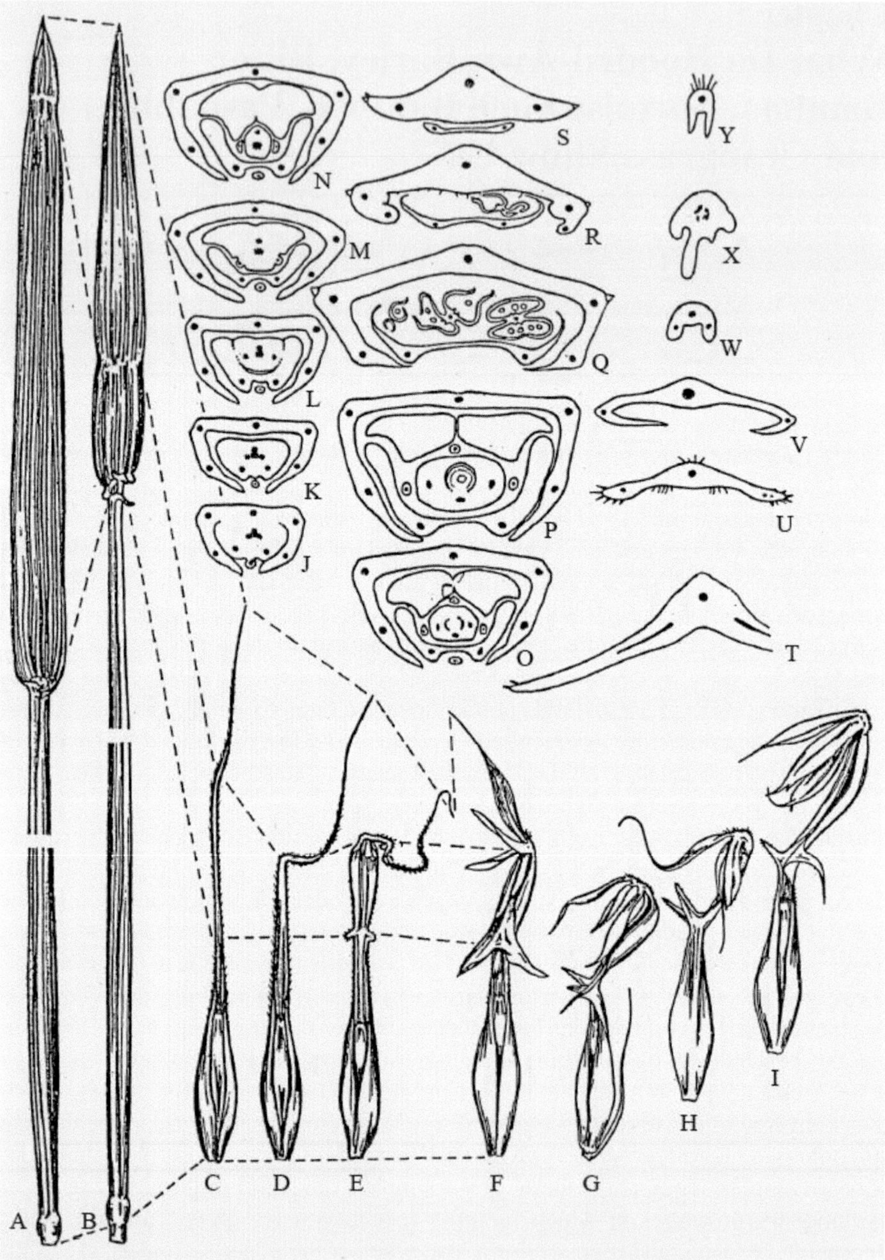

Fig. 5.1 Comparison of the leaf organs of hooded awn barley. (**a**) A leaf grown from the lower part of a stem, showing a constriction on the apical portion of the lamina; (**b**) a flag leaf showing a constriction in the middle or lower portion of the lamina; (**c**) a palea with a normal awn, equivalent to a node at the foliar constriction; (**d**) a hooded palea showing bending at the awn node; (**e**) a leafy palea showing a protoglume spikelet grown from the second axilla; (**f**) two pairs of auricles formed at the node of a leafy awn; (**g**) two supernumerary spikelets grown at the second node facing away from each other; (**h**) two back-facing supernumerary spikelets on the second node merging at their backs; (**i**) two back-facing supernumerary spikelets on the second node merging at their backs and then splitting left and right, forming two auricle-like structures; (**j**–**y**) the transections of a barley floret from bottom to top

stem nodes are all compound nodes formed through the convergence of multiple axes and the disappearance of internodes. On a wheat spike node, the axillary bud of the upper axis develops into a spikelet; it becomes a supernumerary spikelet if the axillary bud of the lower axis also develops into a spikelet.

From the transections of a bitter bamboo (*Pleioblastus amarus*) leaf in Fig. 5.2, we can see four rounds of vascular bundles extending down into the sheath. The outmost round came from the outer ligule; the innermost round came from the inner ligule; and the large second round and the small circled third round came from the lamina on the sheath node (Fig. 5.2c(a)). We can clearly see, in the petiole in Fig. 5.2d(b) and in Fig. 5.2c(a), that the small circled third round of vascular bundles are still very strong at this position, indicating that this round of vascular bundles will soon stop their downward extension, finishing their function in conduction and making room for a second vascular bundle round. However, the third round is stronger than the second at the upper node (Fig. 5.2d(b)–f(c)). These structural changes indicate a turnover relationship among the multiple axes during the organogenesis of the bitter bamboo leaf. As shown in Fig. 5.2f, the axes that build the bitter bamboo's lamina do not split but flatten instead, which is similar to the leaves of garlic chive (*Allium tuberosum* Rottler ex Spreng), which we are to discuss later. From all the axes reveals observed, we can see that nonsplit axes form a round of vascular bundles, while split axes form a row of vascular bundles. In Gramineae and the relative taxa, the downward-extending vascular bundles of the multiple axes constitute multiple vascular bundle rounds or layers. Therefore, the number of rounds of vascular bundles also reflects how many axes are involved in the organogenesis.

As shown in Fig. 5.3, the two constrictions (indicated by the arrowheads in Fig. 5.3a) on the leaf of the tortoise shell bamboo (*Phyllostachys edulis*) reveal that the leaf has been developed by flattening a multiaxial compound node. The long hairs on the outer and inner ligules (Fig. 5.3b–d) belong to metamorphosis at the cellular level, not a structure of an axis. Figure 5.3f shows the vertical and horizontal veins, which the telome theory also finds hard to explain.

In the telome theory, there is an important process of leaf organogenesis, called webbing. That is to say that a web-like structure is grown between the telomes, connecting them together into a lamina. Although this kind of leaf formation from telomes does not exist in nature, webbing, as a process of organogenesis, does. For instance, webbing is a part of the organogenesis that forms spike nodes of paspalum (Fig. 5.4) and crabgrass (Fig. 5.5). In these cases, web tissues connect spike nodes together to form a spike node wing, which extends toward both sides to connect multiple spike nodes into a whole. You can clearly see this kind of structure in a compound palm leaf. Webbing plays an important role in the formation of the main foliar stalk of a compound leaf.

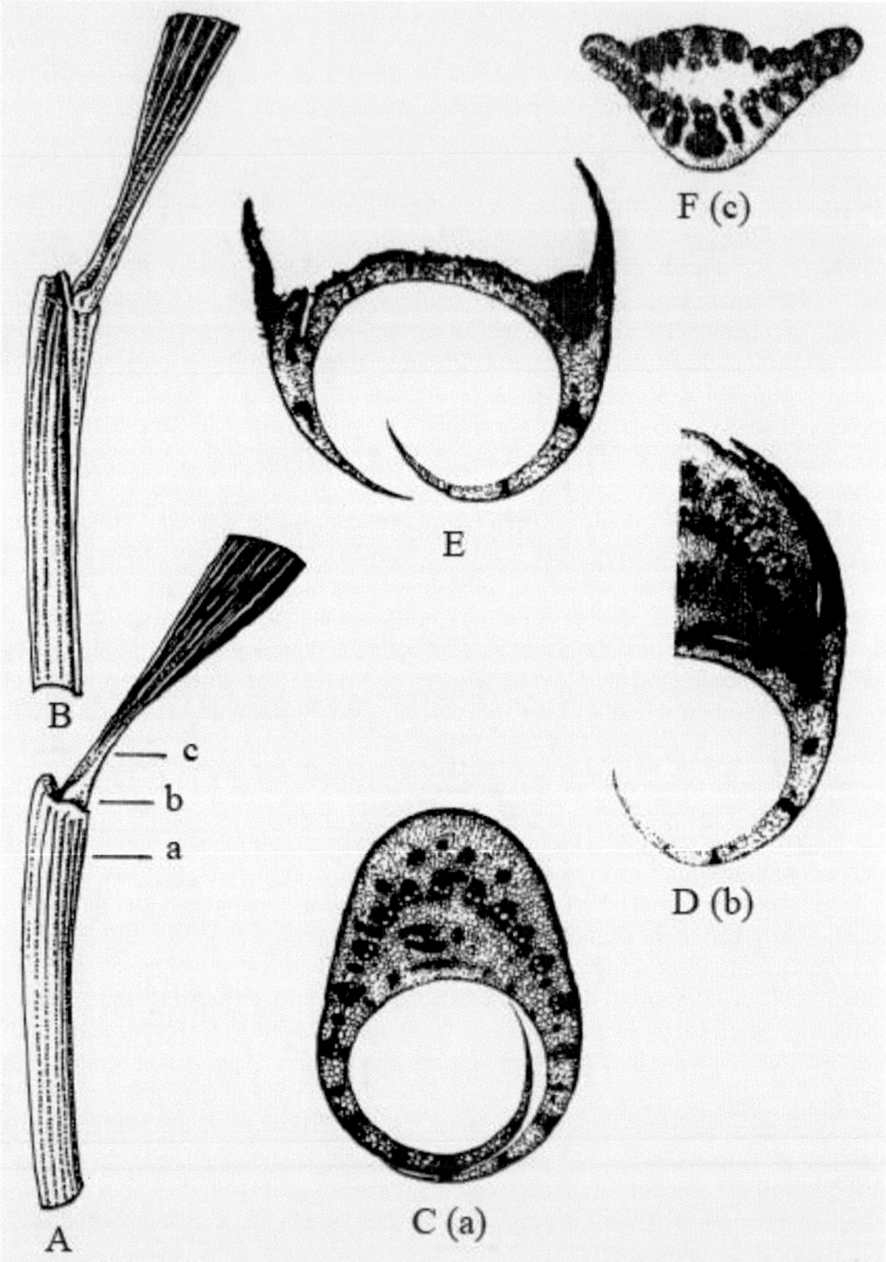

Fig. 5.2 The transections and longitudinal sections of a bitter bamboo leaf, showing the inner and outer ligules and the vascular bundle system. (**a**) The upper portion of sheath and the lower portion of lamina, showing the axillary growth of the inner and outer ligules and the lamina; (**b**) a longitudinal section of the upper section of the sheath and the lower portion of the lamina, showing the axillary growth of the inner and outer ligules and the lamina; (**c**) transection of the sheath showing the inner and outer ligules and three downward-extended vascular bundle systems, the split axis structure of the vascular bundles in the inner and outer ligules, and the close structure of the lamina vascular bundles; (**d**) transection of the conjunction between the lamina and the sheath; (**e**) the same as in (**d**) but without the lamina base; (**f**) transection of the lamina base, showing the closing and splitting of the axis

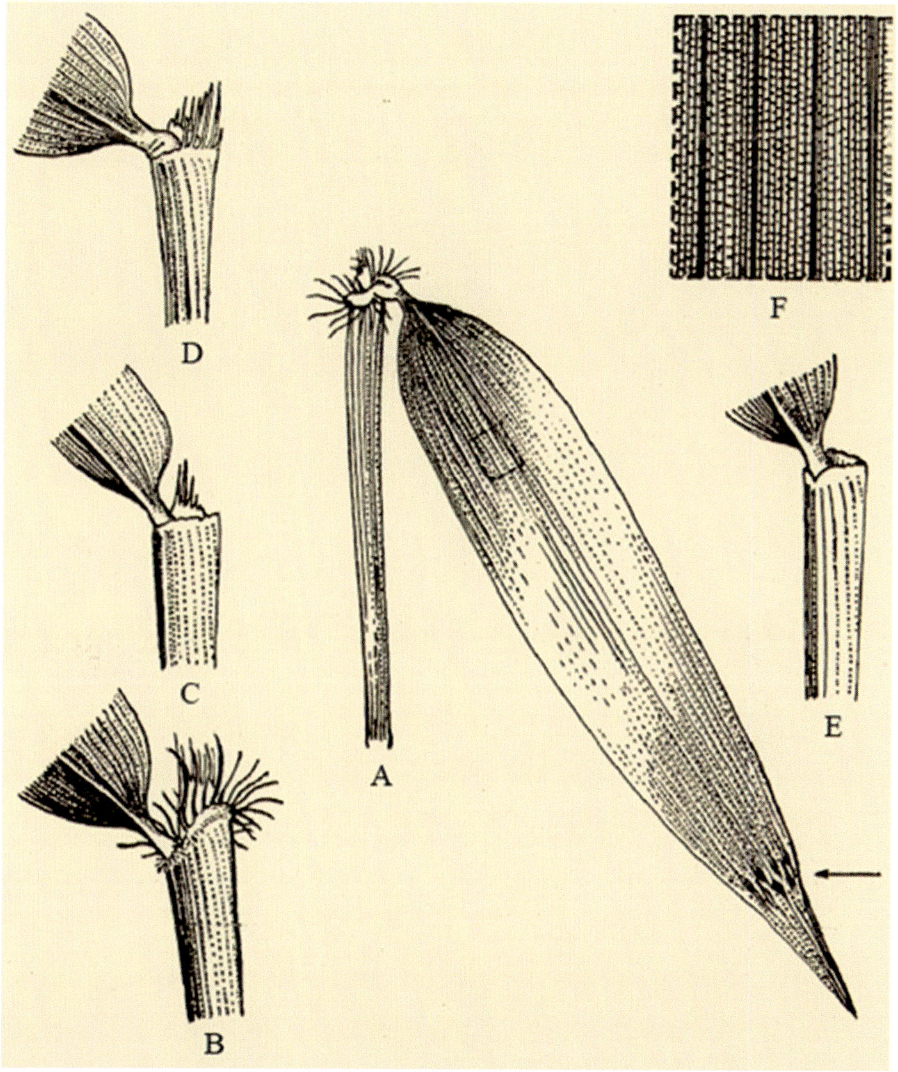

Fig. 5.3 The sheath and lamina of tortoise shell bamboo (*Phyllostachys edulis*), showing the outer and inner ligules and different kinds of hairs. (**a**) The sheath and lamina, the arrowhead pointing to a constriction; (**b**) the inner and outer ligules with long hairs; (**c**) a hairy inner ligule; (**d**) a hairy outer ligule; (**e**) hairless inner and outer ligules; (**f**) a simultaneous vertical and horizontal extension of parallel veins

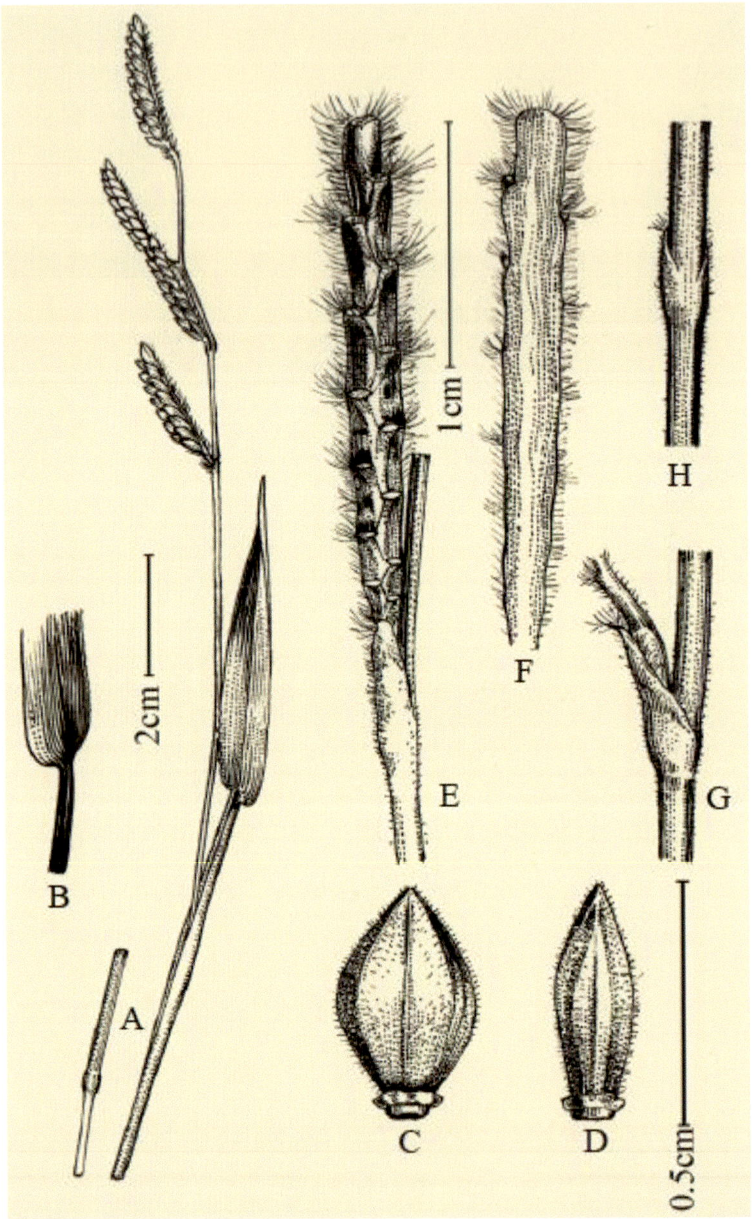

Fig. 5.4 Southern paspalum grass (*Paspalum commersonii* Lam.). (**a**) Spike and flag leaf; (**b**) the middle and lower portions of the flag leaf and the upper portion of the sheath; (**c**) a front view of a spikelet, a side view of a spikelet; (**e**) a ventral view of the spike axis, showing that the spike node and spike internode are webbed together as a wing; (**f**) a dorsal view of the spike axis, showing that the spike node and spike internode are connected together by a webbing wing; (**g**) the base of an axillary branch spike and the hull; (**h**) the convergence of the hull and the internode

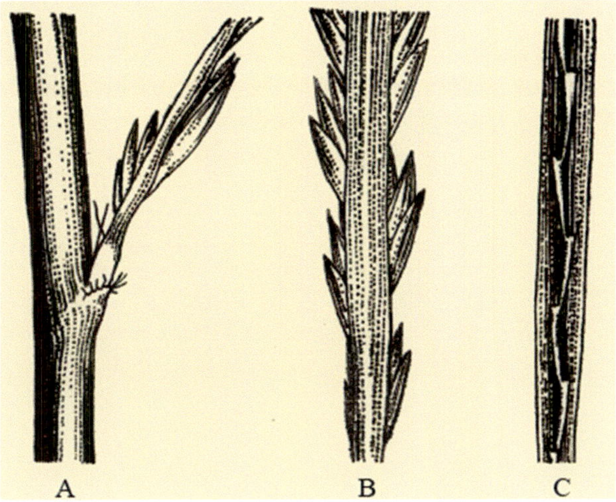

Fig. 5.5 The spike structure of hispid crabgrass (*Digitaria setigera* Roth ex Roem. & Schult.). (**a**) The main spike axis and the lower portion of a branched spike; (**b**) a dorsal view of a branched spike, showing a webbing wing; (**c**) a ventral view of a branched spike (with spikelet removed), showing a webbing wing

Chapter 6
The Ontogenesis of Higher Plants: Pteridophyta

Observations of various ontogenetic processes in higher plants can reveal how organogenesis occurs. These observations can also test if the multiple secondary axis theory is correct or not. Let us first look at Pteridophyta. Most ancient pteridophyte species, such as cuneus-type ferns (Sphenophyllales), are extinct. For the others, there are no specimens in the fossils to show each stage of their whole ontogenesis. Nevertheless, we have evidence to show the whole ontogenesis of current pteridophyte species and, thus, can explain how things happen. Since we have had an in-depth discussion of *Psilotum*, which is close to cuneus-type ferns, it will not be included in our review of each type of current pteridophytes. The pteridophytes that we are going to discuss can be classified into four orders: Filicopsida, Isoetopsida, Lycopsida, and Sphenopsida. They will be reviewed individually as follows.

6.1 Filicopsida

First, let us look at *Angiopteris*'s prothallus on its gametophyte (Fig. 6.1(1)), which has a similar structure to that of Anthocerotae's gametophyte. After fertilization, an egg in the archegonium on the prothallium germinates and grows into a sporophyte. The sporophyte, unlike that of Anthocerotae, is not a parasitic reproductive organ (capsule) on the gametophyte but a nutrient-independent green leaf and a root (Fig. 6.1(2)) formed by splitting and flattening the upper and lower sections of the gametophyte (axis), respectively. This morphogenesis is exactly the same process as how a wheat glume is formed, that is, by splitting the axial structure of a supernumerary spikelet. From the young gametophyte of *Cyrtomium fortune*, shown in Fig. 6.2, we can clearly see that the second leaf grows in the first leaf's axilla and the primordia of the third leaf grow in the second leaf's axilla, which is fully consistent with the evolutionary model of the supernumerary spikelet of wheat. This evolutionary model is based on the multiple secondary axis theory, not the

C. Yen, *Constitution of Organs of the Higher Plants*,
https://doi.org/10.1007/978-981-19-1685-4_6

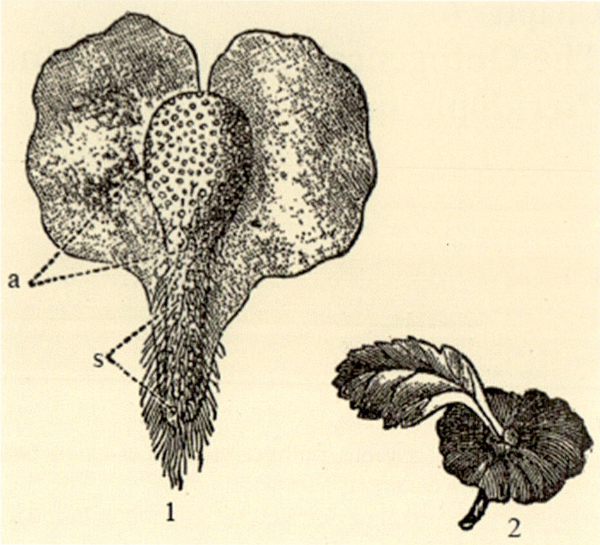

Fig. 6.1 The prothallus of oriental vessel fern (*Angiopteris*). (1) A ventral view, s: antheridium, a: archegonium. (2) A sporophyte growing from a prothallus

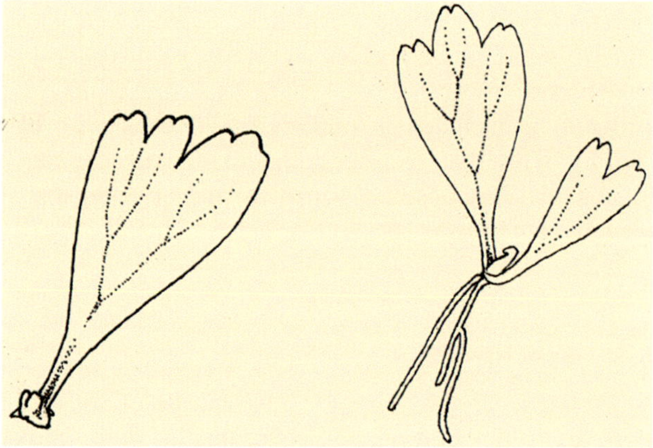

Fig. 6.2 Young sporophytes of holly fern (*Cyrtomium fortunei* J. Sm.), showing axillary growth of leaves

one speculated by the telome theory. The upper portion of a gametophyte splits to become a leaf; an axillary bud (the second axis) from the split petiole splits again to become the second green leaf. The third, the fourth, and all of the upper green leaves form following the same mode. The primordium of the second leaf of *Pteris serrulate* L. also grows from the first leaf's axilla (Fig. 6.3). The sporophytes of *Ophioglossum nudicaule* Christ and *Helminthostachys zeylanica* (L.) Hook. also grow from the axillae of trophophylls (Fig. 6.4). The leaf structure of *Azolla imbricata* (Fig. 6.5) and the configuration of *Salvinia nutans*' leaf primordia (Fig. 6.6) also fit the secondary axis model but not the telome theory.

In Pteridophyta, the vast majority of upper leaves are compound leaves evolved from simple leaves as seen at the lower nodes. The evolution of the compound leaves

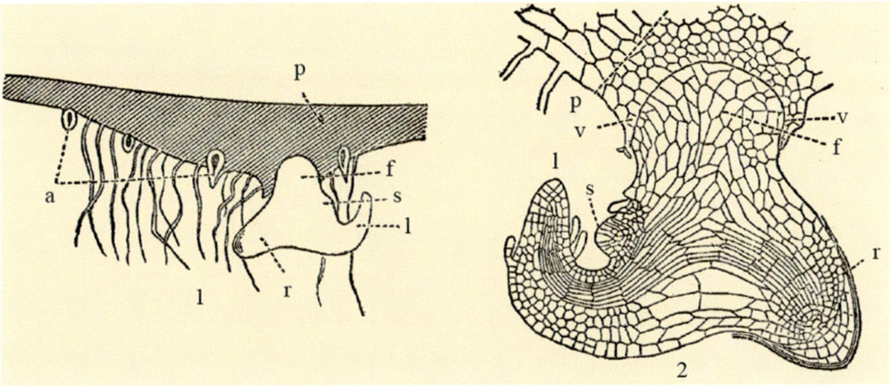

Fig. 6.3 Sporophytes of spider brake (*Pteris serrulata* L.) growing on the gametophyte. (1) A section of the prothallus (gametophyte), showing growing sporophytes underneath; (2) a longitudinal section of the sporophyte—a: archegonium, f: embryo foot (the lower end of the first axis), l: the first leaf (the upper end of the first splitting axis), p: the prothallus, r: the first secondary root, s: the primordium of an axillary axis, v: the inner wall forming on the prothallus through enlargement. (After Figure 295 in Ikeno 1906)

(pinnate or nonpinnate) from simple leaves has gone through a development process wherein layers of secondary axes regenerating on a simple leaf undergo multiple developments, such as splitting, healing, webbing, and converging. Only very few species, such as *Azolla imbricata* (Roxb.) Nakai, maintain simple leaves (scale-like leaflets) at their upper levels.

Let us also look at the anatomy of *Osmunda japonica*. The petiole of its compound leaf is a split axis (Figs. 6.7, 6.8, and 6.9). The stele vascular bundle system is composed of secondary vascular bundles, which are formed by splitting the petiole and extend downward and are connected with each other. This is a typical mode of multiple secondary axes. According to conclusions from physiological experiments, these conducting tissues are formed by the secondary differentiation of the thin-walled cells in the primary tissue under the control of auxin, i.e., the lower ends of the conductive tissues bipolarly differentiated from the lower portions of each of the secondary axillary axes and the vascular bundles of the axes below them interconnecting with each other to form the whole structure of a secondary stem. Figure 6.7 clearly shows the vascular bundle structure of *O. japonica*. Furthermore, the morphology of the leaves and stems of *Cyclosorus aridus*, illustrated in Fig. 6.10, shows such structure much more clearly; i.e., the stems are composed of secondary axes that are connected and fused into each other. If we did not know such a process as that of leaf formation on a wheat supernumerary spikelet by splitting the axis, we would easily believe that, as the phytonic theory states, petioles fuse into each other to form a stem.

According to the telome theory, a vein is an aggregated telome bundle. Dichotomous branches are the original configuration, which evolves into other forms of leaf venation. However, the reticulate venation of *Cyrtomium fortunei*, as illustrated in

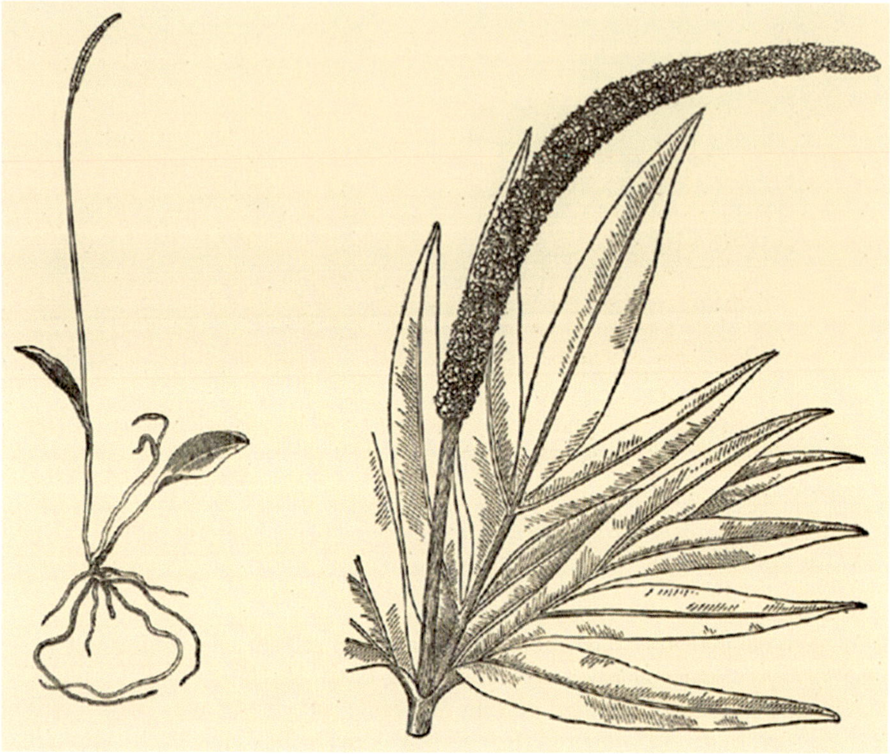

Fig. 6.4 Bottle ear grass (*Ophioglossum thermale* Kom.) (left) and seven-finger fern (*Helminthostachys zeylanica* (L.) Hook.) (right), showing the growth of sporophytes from the axes of a trophophyll. (Adapted from Figures 274 and 277 in Ikeno 1906)

Fig. 6.11, is hard to be explained by the telome theory. From the perspective of ontogenesis, as discussed above, conducting tissues such as veins are differentiated from the thin-walled cells of the primary tissue under the control of auxin. Whether it is a single or an unbranched vein, a dichotomously branched or a reticulate venation, they are all formed from the thin-walled cells along the tracks of auxin flow. That is easy to understand.

In the petioles, different ferns have different vascular bundle structures due to different auxin flow tracks. In *Osmunda japonica*, the vascular bundles transectionally appear to be a broken ring or split cylinder. The dichotomously branched lower portion combines with the dichotomous vascular bundles of other leaf clefts to form the netted central core of the secondary stem structure. Similar broken cylindrical vascular bundles in the petioles of *Cyrtomium fortunei* form the primary single stele in the stem. The petioles of *Cyclosorus aridus* have two parallel vascular bundles, which extend downward into the stem to form the two-column stele by fusing with the vascular bundles from other petioles. The vascular bundles in the petiole of *Cyrtomium fortunei* are divided into three. In the petioles of *Lygodium*

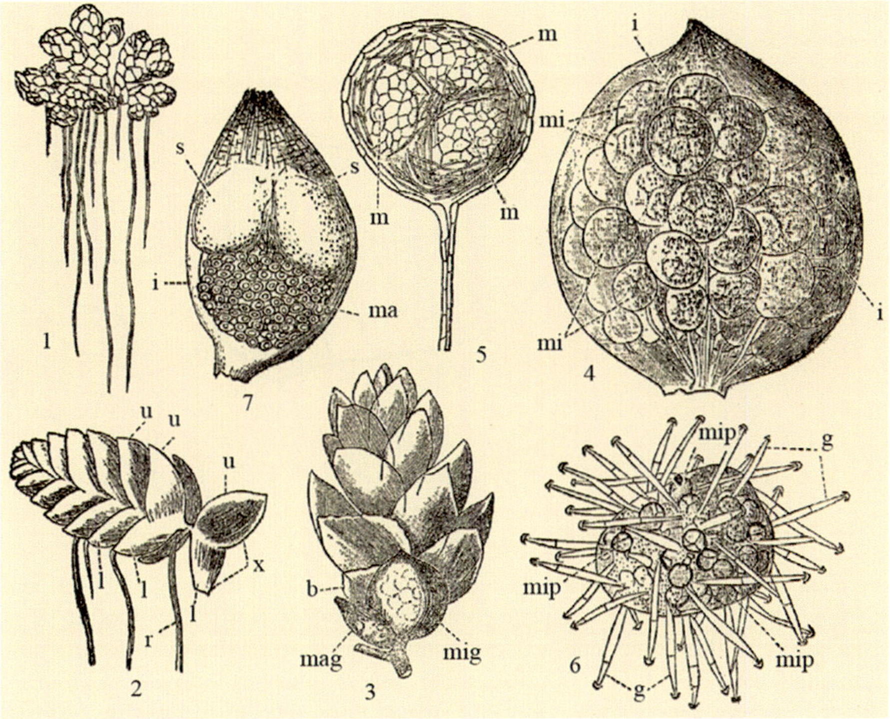

Fig. 6.5 Water velvet (*Azolla imbricata* (Roxb.) Nakai.)—(1) the whole plant; (2) a branch, showing a leaf is split into upper (u) and lower (l) leaf clefts (x: the two leaf clefts were artificially separated to show the splitting); (3) the branches with a megasporangium (mag) and a microsporangium (mig); (4) a microsporangium fruit, showing the coat membrane (i) and the sporangium (mi); (5) a sporangium, showing three spheroids (m); (6) a spheroid, showing the hair (g) and microspore (mip); (7) a megasporangium fruit, showing the megaspores (ma), plankter (s), and coat membrane (i). (After Figure 302 in Ikeno 1906)

japonicum, a single vascular bundle becomes triangular star in shape in the xylem (Fig. 6.12). These vascular bundles are in diverse forms. They are all secondary structures induced by auxin. The observed differences are due to the differences in auxin dosage at particular positions, which are determined genetically.

According to the telome theory, sporangia are on the top of telomes, i.e., at the tips of the veins. However, the vast majority of species in Pteridophyta do not have their sporangia at the tips of their veins. Of them, the sporangia of *Christensenia assamica* (Griff.) Ching are scattered under the leaves. They get their nutrients from the veins via four to six networked microveins (Fig. 6.13). These examples are hard to be explained by the telome theory.

The leaves of the vast majority of Pteridophyta plants are pinnately compound, which are formed through the formation of secondary growing points on the existing leaf clefts, which then develop into secondary axes, which in turn develop into compound leaves. On the plants of *Dicranopteris dichotoma* (Thunb.) Bernh., we

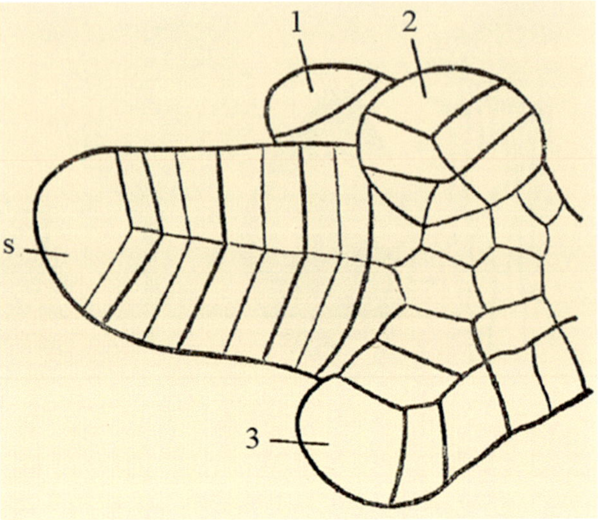

Fig. 6.6 An illustration showing the growing point at the stem top of watermoss (*Salvinia natans*). (1, 2) Aerial leaves; (3) an aquatic leaf, s: a new axillary axis. (After Figure 301 in Ikeno 1906)

can clearly see that a secondary axis grows in the axilla of a young leaf cleft and develops, step by step, into a pinnate compound leaf (Fig. 6.14). The growing point in the axilla of this compound leaf can continue to generate axes and form a new compound leaf (Fig. 6.15). In some other genera, such as *Athyrium viviparum* Christ., *Tectaria fengii* Ching & C.H. Wang, *Polystichum stenophyllum* Christ, and *Woodwardia unigemmata* (Makino) Nakai, these secondary axes can also develop into bulbils for vegetative propagation. The bulbils are specialized shortened and enlarged nutrient-rich branches developed by connecting axes that grow in the axillae of the flattened axes constituting laminas. All these cases of secondary growth are contradictory to the telome theory but are the models that the multiple secondary axis theory is based on.

In Pteridophyta, the compound leaves of *Lygodium japonicum* have reached the extreme of secondary axis development in the higher plants (Fig. 6.16). The continuous development of secondary axes makes the compound leaves become a 1–5-m vine, which look like a vine stem wrapping around shrubs or other attachments. However, the lowermost section of its petiole has two longitudinal ridges (indicated by arrowheads in Fig. 6.12), which divide the petiole into clearly recognizable dorsal and ventral sides, clearly showing that the ridges are the edges of the split axis. If you look at the leaves of all higher plants, no matter how they have changed in shape, the split axis structure never changes. This fully proves that a split axis constitutes a leaf, just like in the model, wherein split axes form the glumes of the supernumerary spikelets in wheat.

Antrophyum obovatum Bak., which is distributed in southwestern China (Yunnan, Guizhou, Sichuan, etc.), does not have a main vein on its leaf blade (Fig. 6.17). All it has are interconnected small reticular veins. The lamina also connects down to the petiole by three to six vascular bundles of similar sizes. This phenomenon is even more difficult to explain by the telome theory.

Fig. 6.7 An illustration showing the vascular bundle system of the leaf clefts of *Osmunda japonica*, which constitutes the stem's stele. Different leaf clefts are shown in different colors

As shown in Fig. 6.8, *Osmunda japonica*'s first leaf is formed from the first axis through its splitting and then flattening into a dichotomous vascular bundle (the vein) on the prothallus (gametophyte). The second leaf is larger than the first one and has tetrachotomous veins. Its blade clearly shows two lobes. The third leaf is even larger with more vein branches. The fourth leaf not only is larger than the third one but also grows a third leaf lobe in the hollow between the first two lobes. The third lobe is obviously formed from a secondary axis by flattening and webbing. On the fifth leaf, the three lobes have evolved into three independent leaflets with ear-like bumps on both sides of the petiole base. A fully developed adult plant has multiple levels of pinnate leaves with ear-like lobes at the petiole bases. The transection of the vascular bundles in *Osmunda japonica*'s petiole shows an open ring, all the way from top to

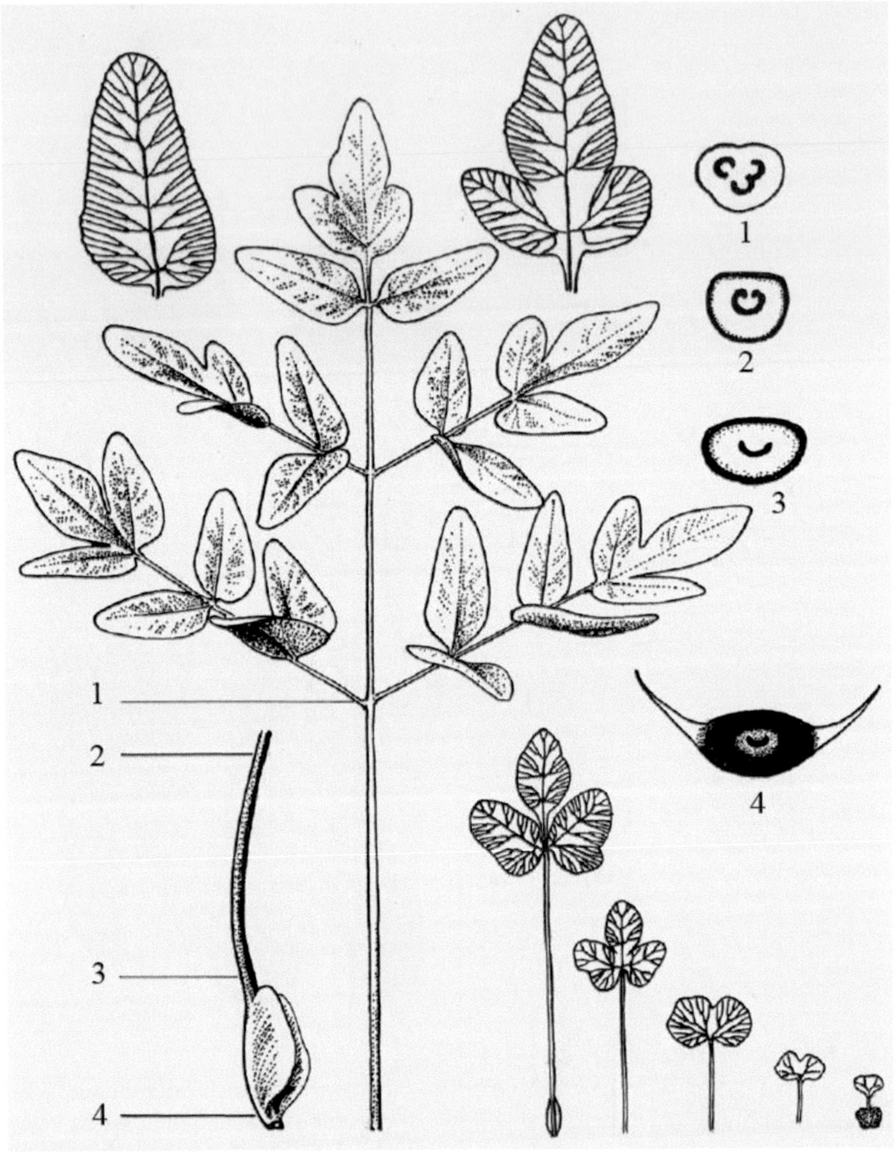

Fig. 6.8 *Osmunda japonica* Thumb., showing the morphology of plants from one with its first young leaf all the way to one with adult leaves. Also shown are petiole vascular bundles with split axial structures

bottom (Figs. 6.8 and 6.9), demonstrating the characteristics of split axes. Although its petioles have been specialized from split axes to become cylindrical, the vascular bundles are still open transectionally, clearly showing the characteristics of split axes.

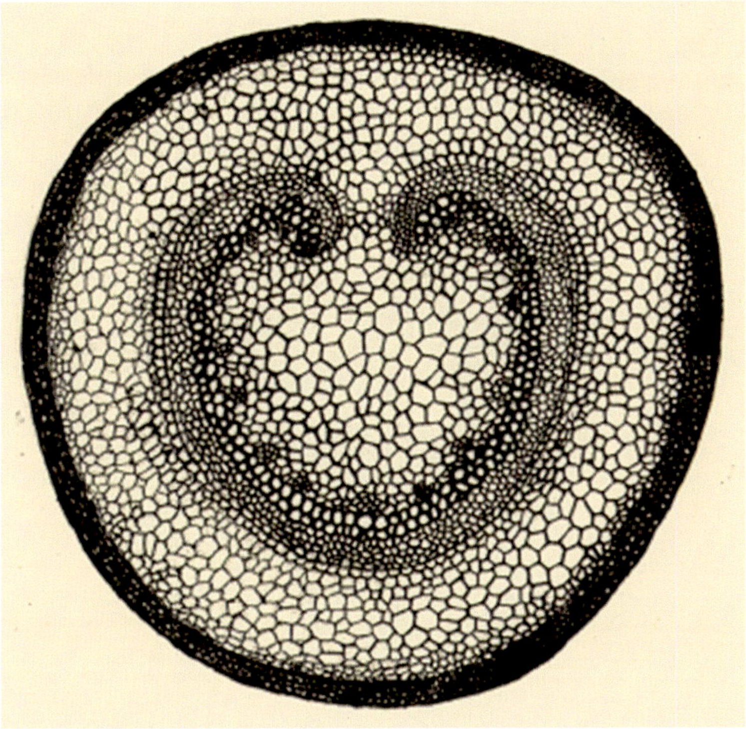

Fig. 6.9 A transection of the petiole of *Osmunda japonica*. Showing vascular bundles and the characteristics of a splitting axis

With the reference to the lemma evolution of hooked barley, it is not difficult to understand how multiple secondary axes evolve into a compound pinnate leaf, like the one of *Osmunda japonica*. The compound pinnate leaves of other ferns should be constructed following the same model.

From Fig. 6.10, we can clearly see that the primary stem of *Cyclosorus aridus* (D. Don) Tagawa, which has not yet developed root, is composed of single split axes fusing together one by one with clearly visible boundaries.

6.2 Isoetopsida

Isoetopsida only has one genus now—*Isoetes* (quillwort). It is an aquatic plant living in swamps. Isoetopsida plants, like those of Filicopsida, have single-split axes in the sporophyte. Unlike Filicopsida plants, however, their axes have evolved into simple microphylls of 5–50 cm long, with a tongue-like secondary axial structure, small ligule, at the bottom. The bottoms of the axes are not merged to form a long, thin stem; instead, they fuse together to form a tuberous structure with 2–3 valvular

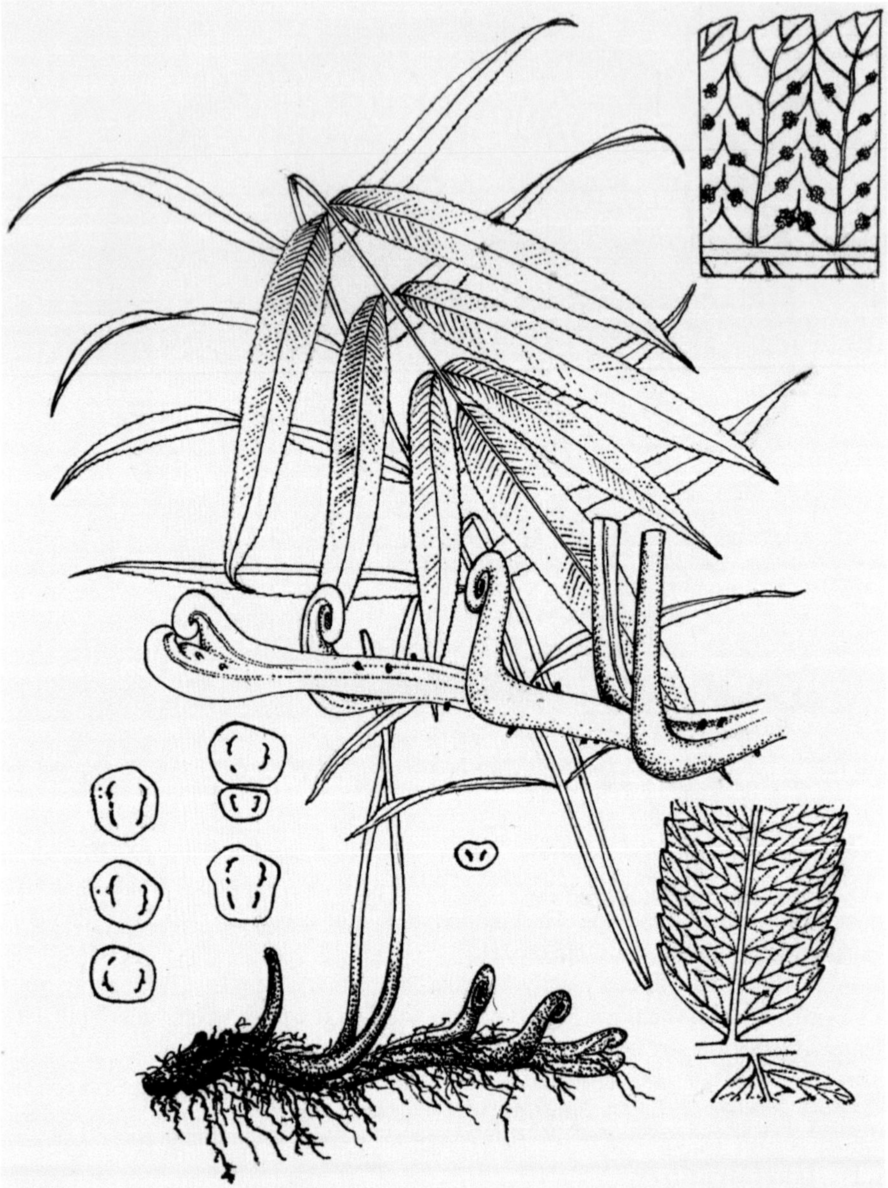

Fig. 6.10 Dry wood fern (*Cyclosorus aridus* (Don) Tagawa), showing the stem forming from leaf clefts by interfusion

dehiscences, like the rhizophore of the Triassic plant *Pleuromeia*. Quillwort's long microphylls gather together on the rhizophore, forming a dense rosette. Quillwort is also unlike lycopods and spike moss (which we will discuss later). Lycopods and

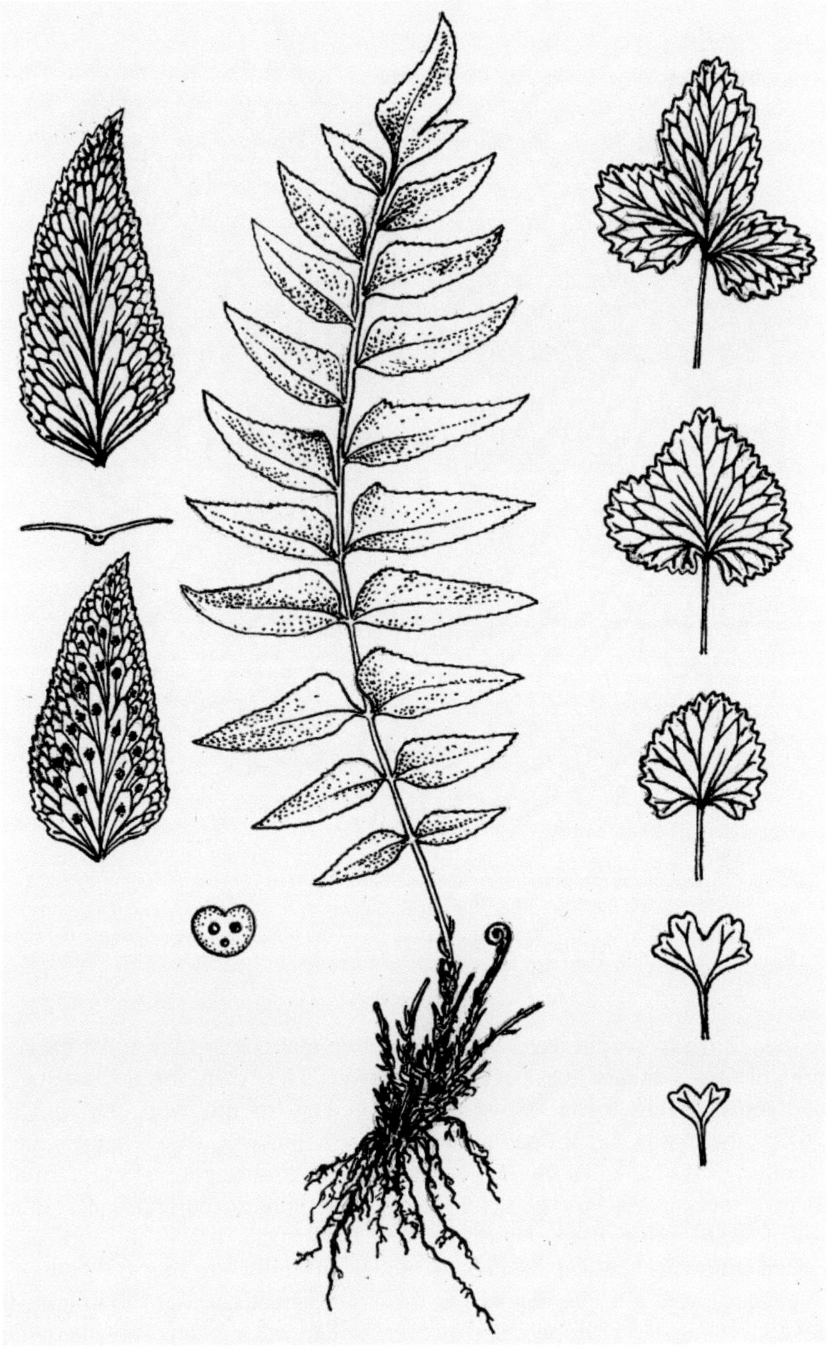

Fig. 6.11 Japanese holly fern (*Cyrtomium fortunei* J. Sm.), showing young green leaves with dichotomously branched veins, adult leaves with mesh-connection veins, and three independent vascular bundles extended downward from a compound leaf into the split axis of the petiole

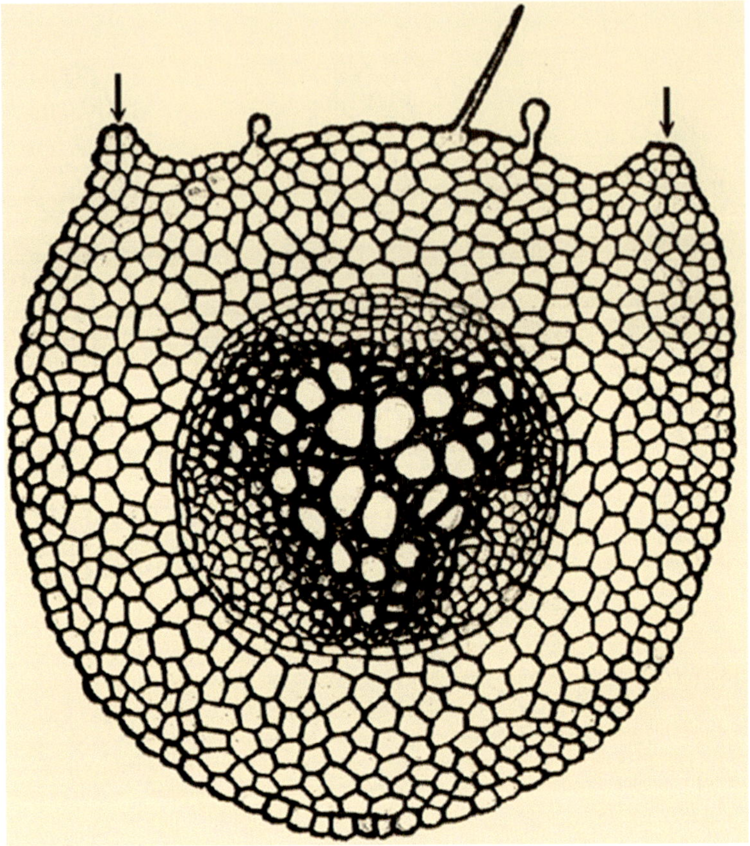

Fig. 6.12 A transection of the petiole of Japanese climbing fern (*Lygodium japonicum* (Thunb.) Sw.). Arrowheads indicate the two ridges, the clear boundaries between the dorsal and ventral sides of the petiole. The splitting occurs during the formation of the blade, and the ridges are the remains of the edges of the splitting axis

spike moss have twin-split axes, which form two leaf clefts. Like spike moss, the splitting of the secondary axes occurs on quillwort's leaf cleft, forming tongue-like small ligules. Figure 6.18a shows the morphology of quillwort. As shown in Fig. 6.18f, the first leaf is formed at the top of the first axis, which is the structure that Smith (1955) called cotyledon, and an embryo foot at the bottom. Therefore, quillwort's roots are the base plate for the aggregation of its multiple axes; i.e., new secondary roots grow from the rhizophore. Quillwort's organ structure is still organized following the secondary axis model. In quillwort, after the bottoms of the secondary axes are interconnected, their meristem tissues at the bottom fuse together, forming the rhizophore. However, unlike other plants, no internode is formed for intervening growth, and thus no stem is developed through elongation. Instead, the leaves of the upper and lower sections gather together, forming a dense

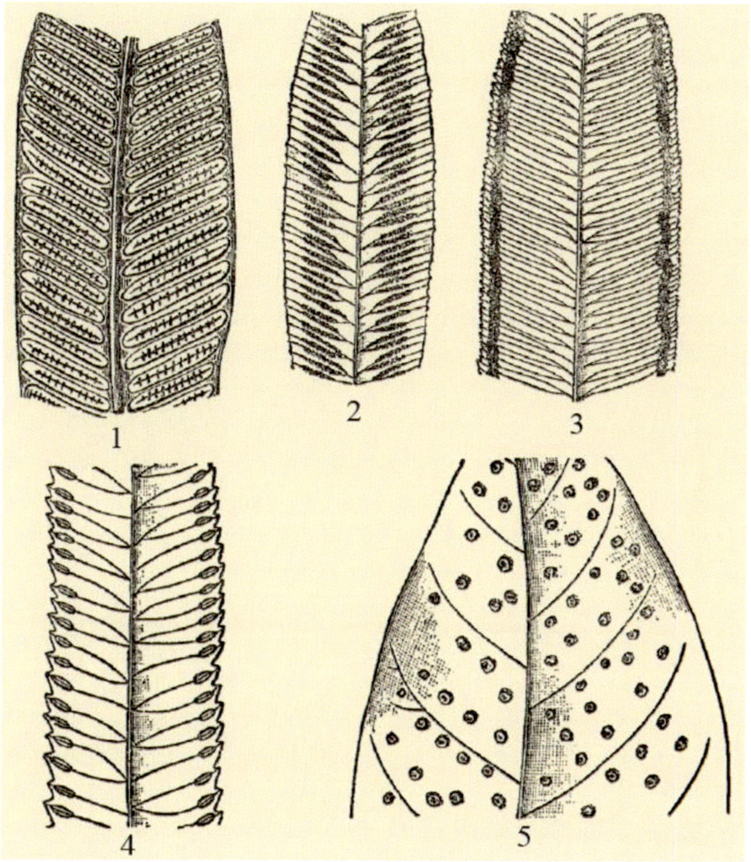

Fig. 6.13 Illustrations of the positions of veins and sporangia of Angiopteridaceae plants, showing sporangia scattering among the veins. (1) *Danaea elliptica*, (2) *Archangiopteris henryi*, (3) *Angiopteris crassipes*, (4) *Marattia fraxinea*, (5) *Christensenia assamica*. (After Figure 271 in Ikeno 1906)

rosette structure. Secondary roots grow from the rhizophore, developing lots of fibrous roots. This is the typic morphogenesis of quillwort's organs.

6.3 Lycopsida

Modern lycopod plants have two large groups: the lycopod order (Lycopodiales) and the spike moss order (Selaginellales). Unlike quillwort (Isoetopsida), lycopod and spike moss plants have dichotomous axes, forming a pair of leaves, while plants in the fern class (Filicopsida) and the quillwort class (Isoetopsida) have simple axes, forming one leaf. Plants of the horsetail class (Equisetopsida) have multichotomous

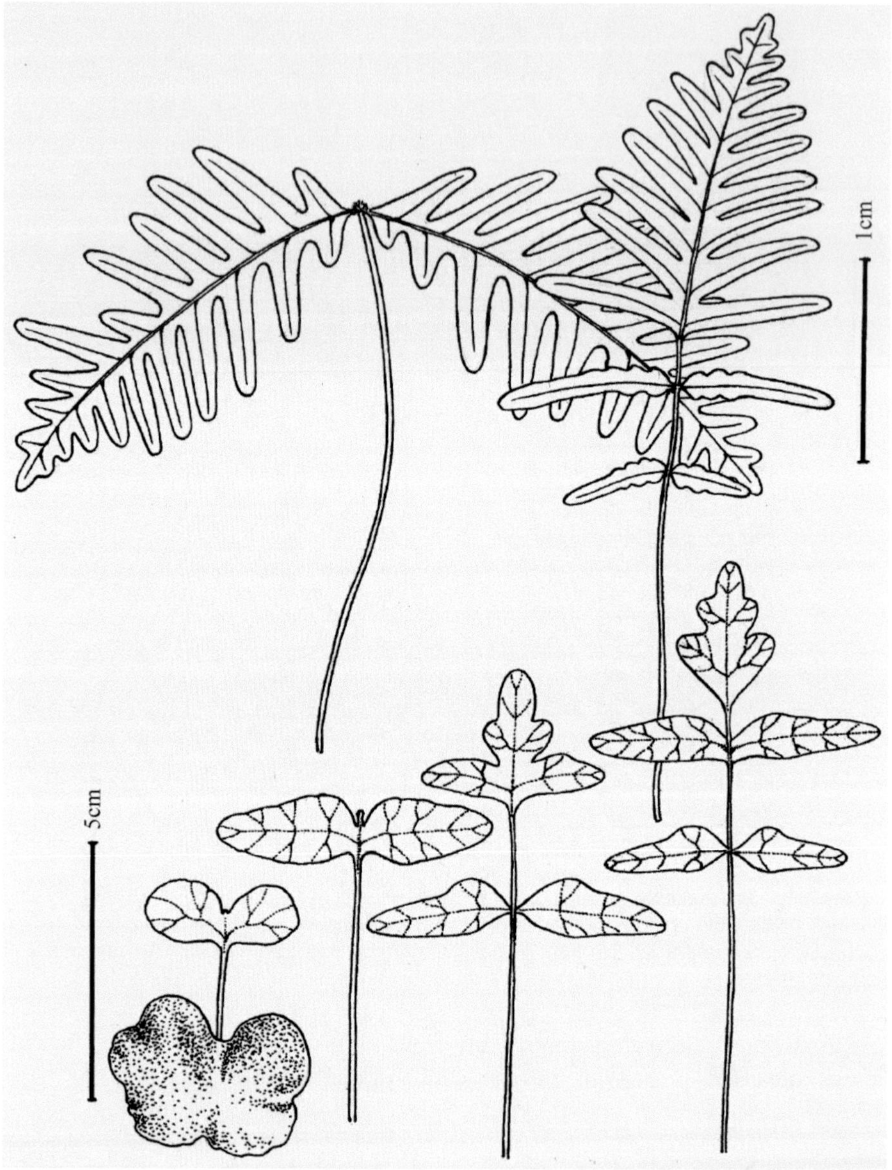

Fig. 6.14 Old World forked fern [*Dicranopteris dichotoma* (Thunb.) Bernh.], showing the different developmental stages of young leaves. Compound pinnate leaves are evolved from multiple secondary axes

axes and form verticillate leaves. Lycopods do not grow complex secondary axis on their leaves. Instead, they keep the original structure of relatively simple leaf clefts, i.e., the microphyll structure. The ferns develop macrophylls with complex

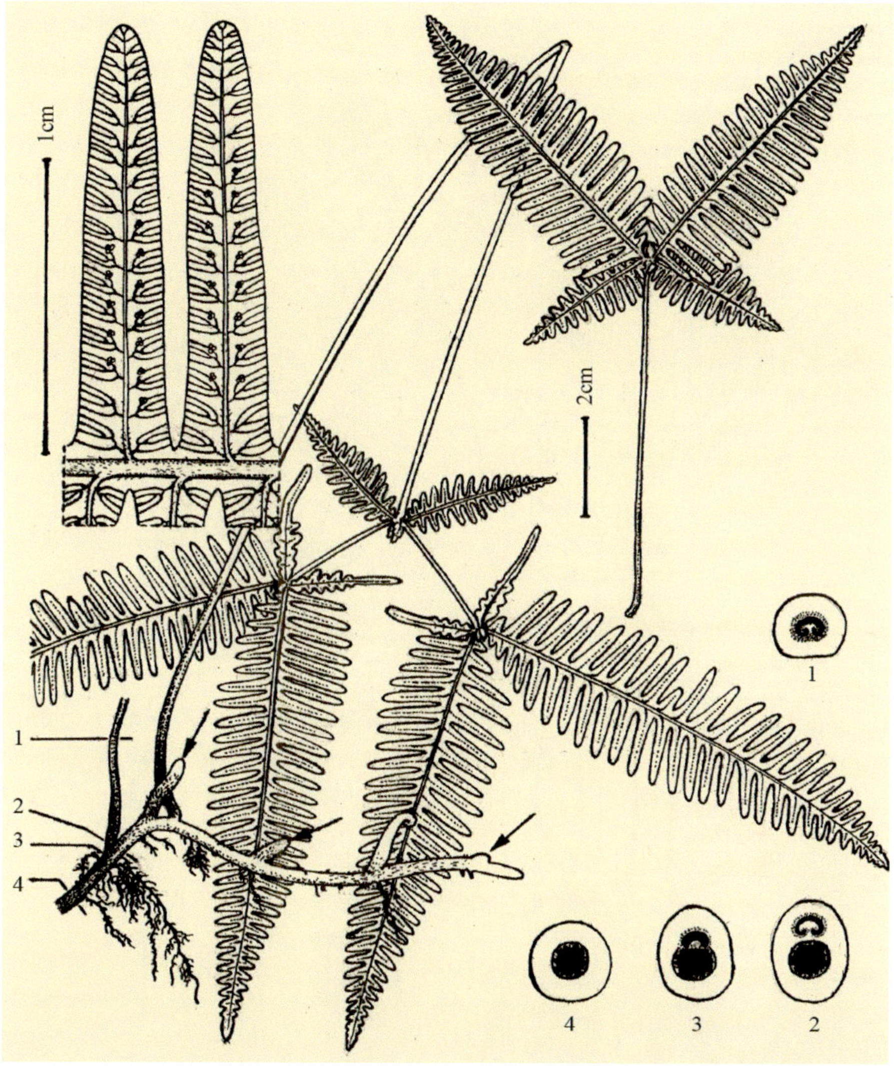

Fig. 6.15 Old World forked fern (*Dioranopteris dichotoma* (Thubb.) Bernh.), showing compound leaves, venations, the splitting vascular bundles in the petiole, and the fusion of the compound stele formed from multiple secondary axes. The arrowheads indicate the ascending secondary axillary buds

structures from multiple secondary axes. The multiple leaf clefts in horsetails have degenerated into scale leaves, with photosynthesis taking place in the stem axes, which are formed at the lower portion of the axes.

Fig. 6.16 Japanese climbing fern (*Lygodium japonicum* (Thunb.) Sw.). (**a**) A prothallium and the first sporangial leaf growing on it; (**b**) the second leaf, showing a compound leaf evolved from a secondary axis on the leaf cleft; (**c**, **d**) a compound leaf forming from more secondary axes; (**e**) the adult plant, showing the development of the vine from a compound leaf; the arrowhead indicates the formation of the vine from an axillary bud (secondary axis); (**f**) sporangia

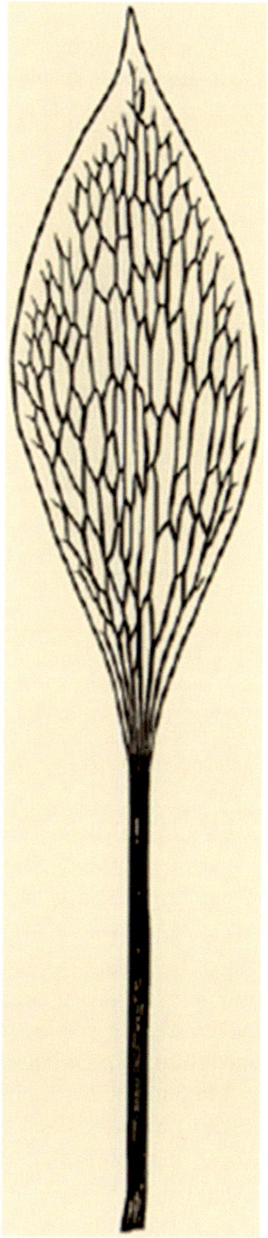

Fig. 6.17 An illustration of a leaf of long-stalk *Antrophyum* (*Antrophyum obovatum*), showing interconnected reticular veins

6.3.1 Lycopodiales

Someone has now classified plants in the lycopod order into two families, Huperziaceae and Lycopodiaceae, on the basis of whether their stems are creeping

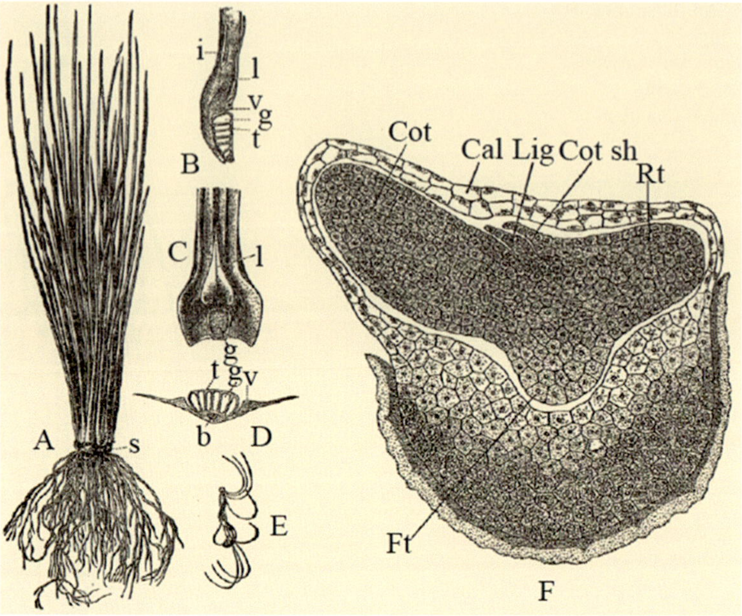

Fig. 6.18 Morphogenesis of quillwort (*Isoetes*). (**a–d**) *Isoetes japonica*. (**a**) The whole plant, (**b**) the longitudinal section of the base of a sporophyll, (**c**) the bottom view of the sporophyll's base, (**d**) a transection of a sporophyll and sporangium (b: vascular bundle, g: sporangium, i: the space between the tissues within the leaf, 1: microphyll, t: grid structure, v: ligule), (**e**) sperms of *Isoetes malinvernianum*, (**f**) a young embryo of *Isoetes braunii* Dur. (Cal: calyptra, Cot: cotyledon, Cot sh: cotyledonary sheath, Ft: foot, Lig: ligule, Rt: root, St: stem). ((**a–d**) after Figure 336 in Ikeno 1906; (**f**) after Figure 142 in Smith 1955)

or upright, whether they develop spike-like sporangia or not, and whether their fertile and sterile sporophylls have the same color and shape or different color and shape. In terms of organogenesis, these characteristics have nothing to do with the principles of organ construction. You can artificially classify these plants into the two classes. However, from an organogenetic viewpoint, both can be classified into one group: *Lycopodium*.

Prothalli in *Lycopodium* do not differentiate between macro (female) or micro (male) prothalli. Each prothallus bears both antheridia and archegonia. In other words, it is both male and female. The sexual generation most often lives underground symbiotically with fungi. When exposed to the surface, it becomes a green vegetative form living on its own. It is large with 0.5–1 cm in length. Figure 6.19 shows a prothallus of *Lycopodium complanatum* L., which differs greatly from that of Filicopsida. On the prothallus, a young embryo (sporophyte) becomes blocky after the fusion of the sperm and egg, forming two protophylls, which Smith (1955) called cotyledon. From leaf axilla, new growing points, the secondary axes, emerge (Fig. 6.20), which develop into the second pair of microphylls.

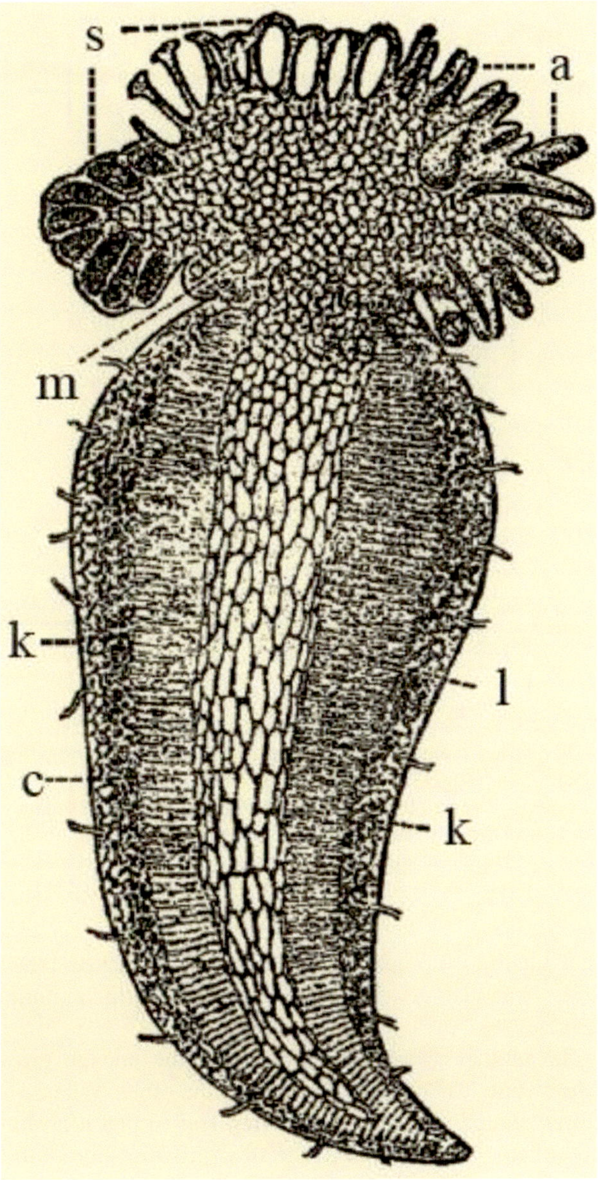

Fig. 6.19 A longitudinal view of a prothallus of *Lycopodium complanatum*—a: archegonium, s: antheridium, m: meristem, c: core, k: epidermis, l: palisade. (After Figure 321 in Ikeno 1906)

As shown in Figs. 6.21 and 6.22, *Lycopodium serratum* Thunb. often develops a kind of short branch for vegetative propagation. Such a short branch has three layers of opposite leaves formed from three levels of axes. A bulbil develops from the axilla between a pair of microphylls formed by axis splitting. The third axis splits to form a pair of microphylls, and a bulbil then develops in the axilla. The bulbil has three layers of opposite leaves and an undeveloped growing point inside, which means

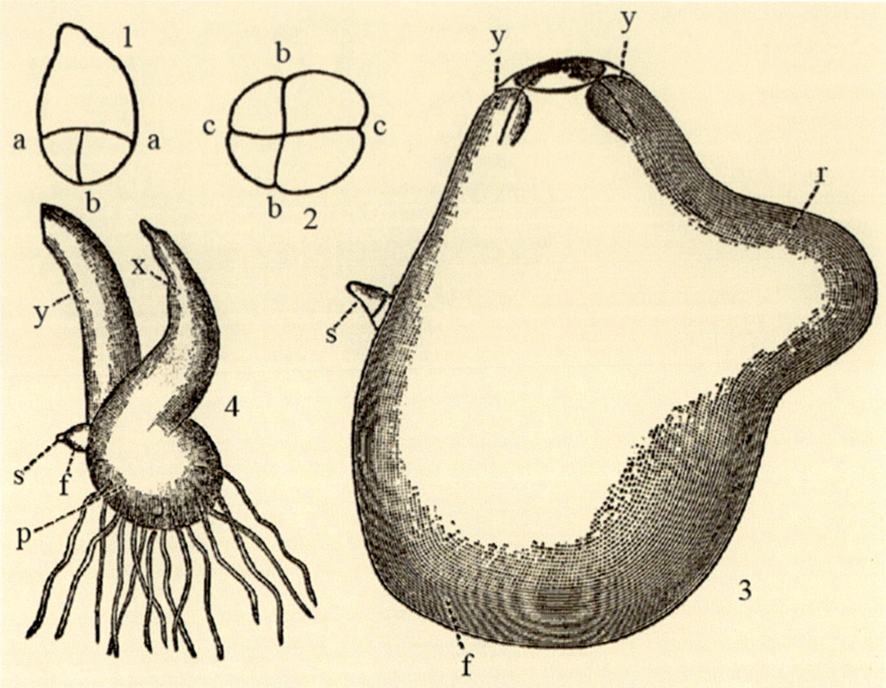

Fig. 6.20 Development of a fertilized lycopod egg. (1) The side view of the fertilized egg in the early division stage—a-a: the portion above the transection will develop into a suspensor; (2) the ventral view of the fertilized egg in the early division stage: dividing twice toward the b-b and c-c directions to form four cells, which will develop into a young embryo; (3) a further developed young embryo—f: foot, r: root, s: suspensor, y: leaf primordia; (4) a further developed young embryo—f: foot, p: the unsplit lower portion of the first axis, s: suspensor, x and y: two young leaves developed via splitting the upper portion of the first axis. (After Figure 322 in Ikeno 1906)

that a bulbil is composed of four layers of stacked axes. After the bulbil roots in the soil, it can also germinate into a seedling, which continues forming opposite green leaves.

In another group of *Lycopodium*, the uneven growth of two opposite leaves causes one leaf to grow faster than the other. When a vascular bundle has already differentiated in the faster-growing leaf so that it no longer elongates and becomes solidified, the lower end of the slow-growing side of the axis remains in a meristem status and continues to grow. When the slow-growing side is also differentiated and matures, the phenomenon that one leaf cleft is taller than the other appears on the stem (the united axes). The originally opposite leaf clefts now become alternate. If the interconnected upper axes have also evolved the same way, the uneven development between the interconnected upper and lower axes must become different in size, causing an alignment shaft on the plane. Therefore, the secondary stem, which is interconnected axes, grows by an ascendingly twisted rotation, making the opposite phyllotaxis gradually a rotating alternate one. These phenotypes clearly

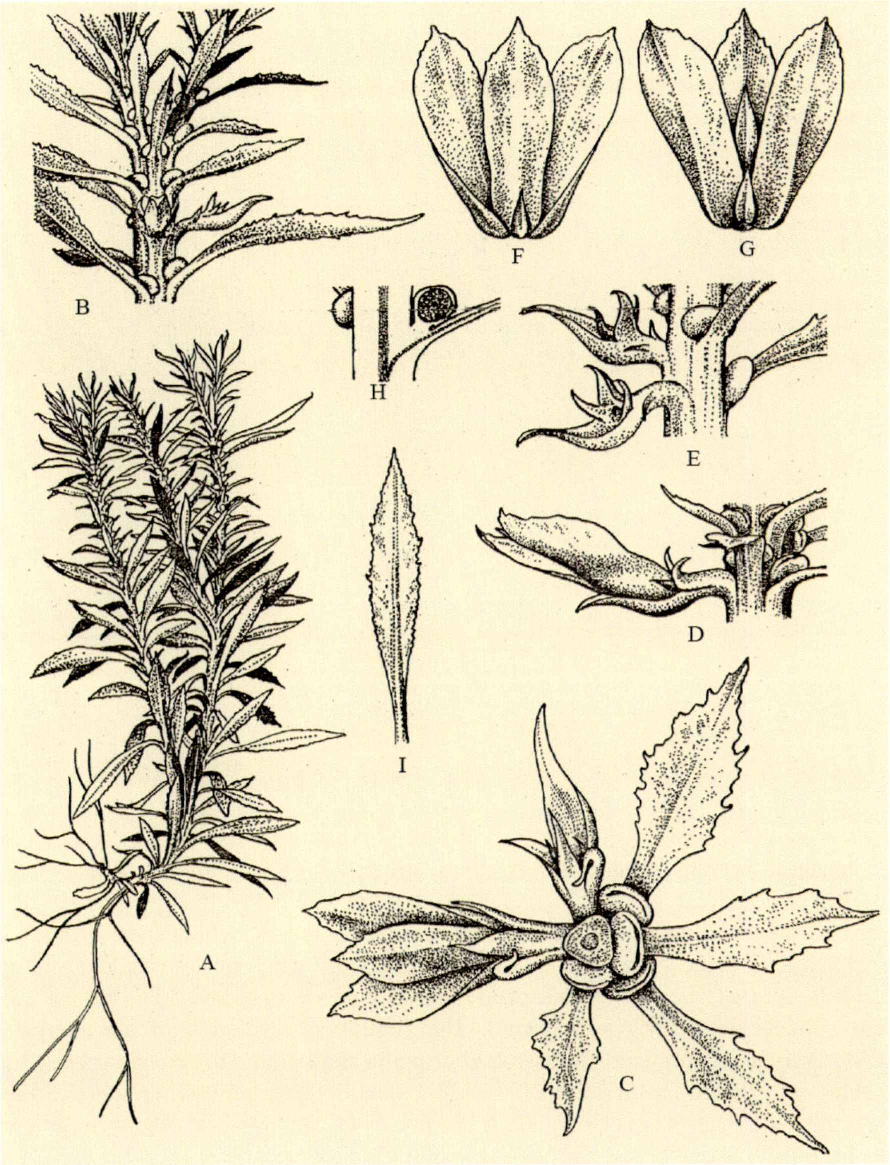

Fig. 6.21 *Lycopodium serratum* Thumb. (**a**) The whole plant; (**b**) a part of the plant showing the vegetative reproduction branch and the sporangia; (**c**) a transverse section of the plant viewed from the top, showing nutrient leaves, sporophylls, vegetative reproduction branches with two secondary axes and four opposite leaves, and bulbils; (**d**) a vegetative reproduction branch with bulbils; (**e**) a vegetative reproduction branch with two secondary axes and four leaf clefts, showing that a bulbil has fallen off; (**f**) a bulbil viewed from the bottom; (**g**) a bulbil viewed from the top; (**h**) a longitudinal view of a stem and sporophyll, showing the fusion of the vein and the stele, and a sporangium at the axilla; (**i**) a trophophyll

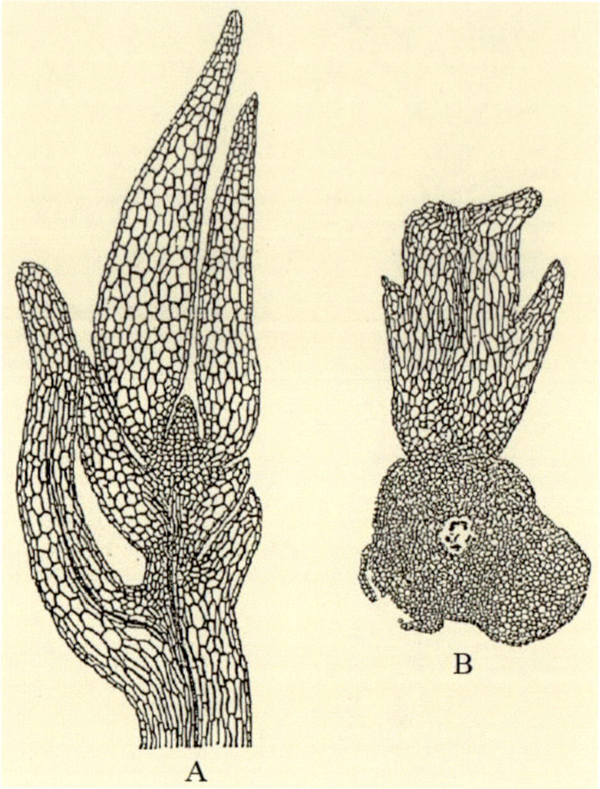

Fig. 6.22 A longitudinal section (**a**) and a transection (**b**) of a vegetative reproduction branch of *Lycopodium serratum*, showing the axillary nature of the secondary axes and the opposite leaves formed by splitting in the upper portion of the axes, which clearly demonstrates the splitting nature of the leaves and the complexity of the stem

demonstrate that organogenesis in lycopods still follows the multiple secondary axis model. The alternate leaves of an adult plant are evolved from the opposite phyllotaxis of the seedling. First, there are two opposite leaves, which then evolve to alternate leaves due to the uneven growth of the pair, which makes one leaf rotates up. It is not that overtopping alternate telomes develop first, and they then evolve into opposite leaves via aggregation. The continuous evolution of the alternate phyllotaxis inevitably leads to its rotation due to the volume difference caused by the uneven growth. The rotation will be left-handed if the upper leaves mature earlier than the lower ones; conversely, it will become right-handed when otherwise. These are the objective reality and can be seen in many lycopod plants, such as *Lycopodium serratum* Thunb., *L. clavatum* L., and *L. cernuum* L., and these are exactly negations of the telome theory. In other words, the examples seen in lycopod plants reject the telome theory and confirm the multiple secondary axis theory.

There are many types of lycopod plants. They are morphologically different in the prothallus of their sexual generation and in the stems, leaves, and sporophyll branches of their vegetative generation. As shown in Fig. 6.21, the leaf of *Lycopodium serratum* Thunb. has a broad lamina with a lanceolate shape; *Lycopodium complanatum* L. has diamond-shaped microphylls, formed by two opposite splits, in four rows along the stem; the leaves of *L. clavatum* L. are small and fine with alternately rotating short hairs. Nevertheless, these leaves are all simple microphylls, with each being formed from an axis by two splits.

6.3.2 *Selaginellales*

Selaginellales (spike moss) is closely related to Lycopodiales but different from the latter in that spike moss has macroprothalli and microprothalli (Fig. 6.23). A macroprothallus only develops archegonia and thus is female-only, while a microprothallus only develops antheridia and thus is male-only.

After fertilization, a macrospore germinates and develops into a sporophyte with opposite leaves (Fig. 6.24). The first opposite leaves on the seedling are simple microphylls. The stem of a mature plant usually creeps on the ground. The four rows of opposite leaves, which are formed by splitting the upper and lower axes, all face upward toward the light source. The upper leaf of the two leaf clefts on each node is smaller and lies flatly along the stem, while the lower one is larger and lies horizontally outward and vertically to the stem. Because the stem is creeping on the ground, the four rows of opposite leaves have adapted to the direction of light during the evolution and, therefore, changed their sizes, shapes, and orientations (Figs. 6.25 and 6.26). The microphylls of spike moss are different from those of lycopods in that they grow small paraphylls on their leaves from axillary axes (Fig. 6.25(2)). Also, from the axillary axes, macrosporangia and microsporangia develop, which further develop into macrosporophylls and microsporophylls, respectively (Figs. 6.25(3), (4)). Figure 6.26 clearly shows that a single secondarily developed vascular bundle in a microphyll extends downward into the stem, which is formed by the aggregation of the upper and lower axes. The secondary vascular bundles converge in the stem, forming a simple primeval stele. They can also

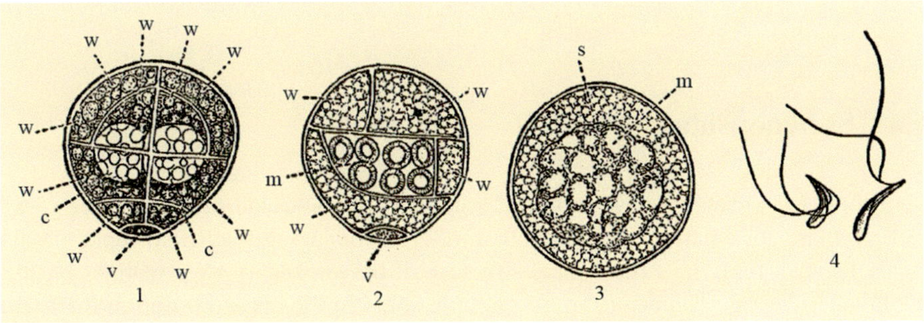

Fig. 6.23 Microsporophytes of *Selaginella*. (1) A microsporophyte (male prothallus) of *Selaginella martensii*—c: centrocytes, v: trophocytes, w: antheridial parietal cells. (2, 3) Microsporophytes of *S. stolonifera* with multiple spermatocytes—m: spermatocytes, s: the slime layer developed from parietal cells, v: trophocytes, w: parietal cells. (4) Sperms. (After Figure 326 in Ikeno 1906)

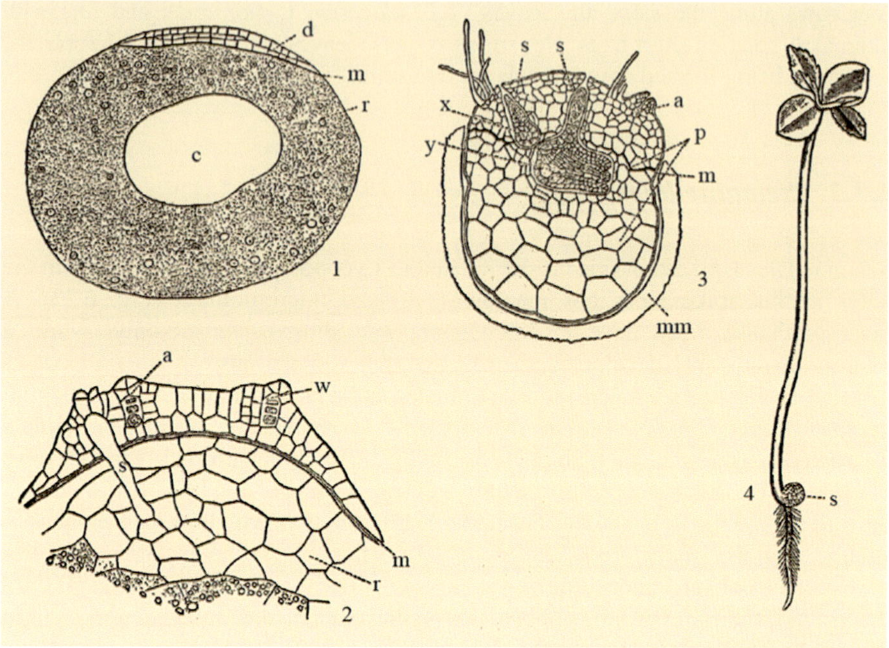

Fig. 6.24 The macrosporophyte and a seedling of *Selaginella*. (1, 2) Macrosporophytes (female prothalli) of *Selaginella kraussiana*—a: archegonium, c: central cavity, d: disk, m: phrenic membrane, r: storage tissue. (3) Macrosporophytes (female prothalli) of *S. martensii*—a: archegonium, m: phrenic membrane, mm: macrospore membrane, p: prothalli, s: suspensor, x: a young embryo, y: a larger young embryo. (4) A sporophyte seedling showing three pairs (three secondary axes) of opposite leaves with one pair being too young to open up; s: macrospore. (After Figures 327 and 328 in Ikeno 1906)

develop into a multiring tubular or multibody stele (Fig. 6.27). Lycopods, on the other hand, usually form star-shaped exarch steles.

6.4 Sphenopsida

Sphenopsida (horsetails) have a typical secondary axial structure. Because their leaf clefts remain in a native state without further secondary growth into lepidophylls (scaly leaves), they have the typical structure of the secondary axes without further evolution. They are characterized by multiple splits in the upper portion of the axes, which develop into cyclic phyllotaxy. This phenomenon is the most direct and obvious negation of the telome theory. The telome theory cannot explain this at all. That is why Smith (1955) said that the origin of horsetails' cyclic phyllotaxy was still unclear.

The prothallus of horsetail's sexual generation is very similar to that of moss, it being male and female (Fig. 6.28). Basically, all of the morphological structures of

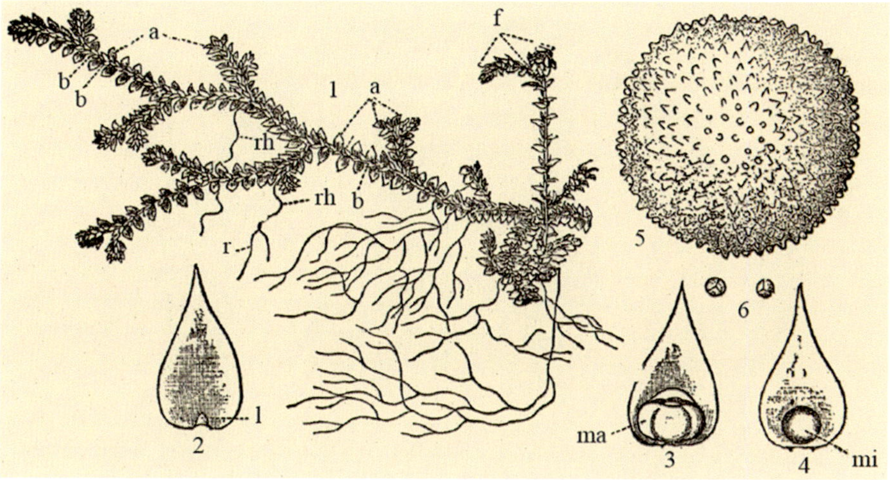

Fig. 6.25 *Selaginella savatieri*. (1) The whole plant—a: an outward-open macrophyll under the stem, b: a microphyll over and close to the stem, f: sporophylls, rh: roots, r: rootlets. (2) A ventral view of the leaf—1: small ligule. (3) A macrosporophyll—ma: megasporangia. (4) A microsporophyll. (5) A macrospore. (6) Microspores. (After Figure 325 in Ikeno 1906)

Fig. 6.26 An illustration of *Selaginella savatieri*, showing the downward connection of the single vascular bundle with the stele

horsetail can be found in moss, that is to say, the genetic characteristics that make up horsetails have already formed in mosses. It can be said that horsetails have a much closer relationship with Jungermanniales of Hepaticae than with Anthocerotales.

The multiple splits at the top of a horsetail's node (Fig. 6.29) are morphologically very similar to a bud dish of liverwort (*Marchantia polymorpha*) (Fig. 6.30). The

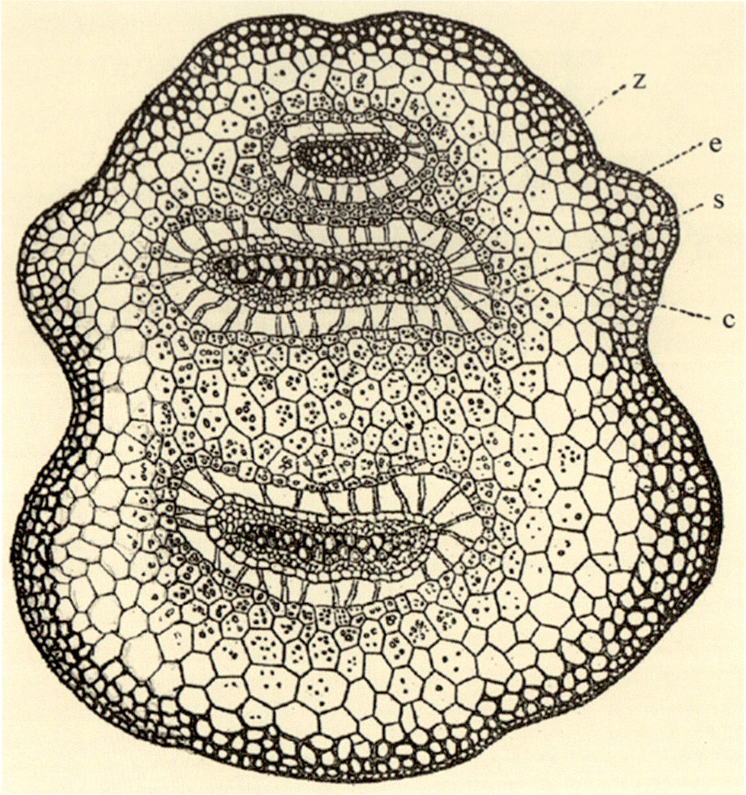

Fig. 6.27 The transverse section of the stem of *Selaginella inaequifolia*, showing multiple steles; c: cortex, e: epidermis, s: air cavity, z: stele. (Modified from Figure 324 in Ikeno 1906)

shape observed in horsetails can be formed if the buds at the inside bottom of the bud dish repeatedly produce the dish structure. Horsetails' strobili can be formed if the moss' umbrella-like sporangia aggregate together (Figs. 6.31 and 6.32). Although the similar morphological structures between them are expressed in different generations, they can also be expressed in the same generation. The bud dish of liverwort is of sexual generation; the similar structure in horsetails is of vegetative generation. The generations are different, but the morphology is similar. It demonstrates that the genetic effects on organogenesis are the same in different generations. The genes are similar. These genes and the organogenesis of horsetails have already appeared in mosses.

Figure 6.29 is taken from Cryptogamic Botany published by Gilbert M. Smith in 1938. It shows the developmental stages of a young embryo that forms in an archegonium on a female prothallus after fertilization. Figure 6.29g shows the longitudinal section of the young embryo. Its calyptron looks like liverwort's bud dish. The young embryo is formed by fertilization of egg and is not derived from the calyptron tissue of the sexual generation. However, the propagation buds in the bud dish are derived from tissues of the sexual generation. The three levels of axes shown

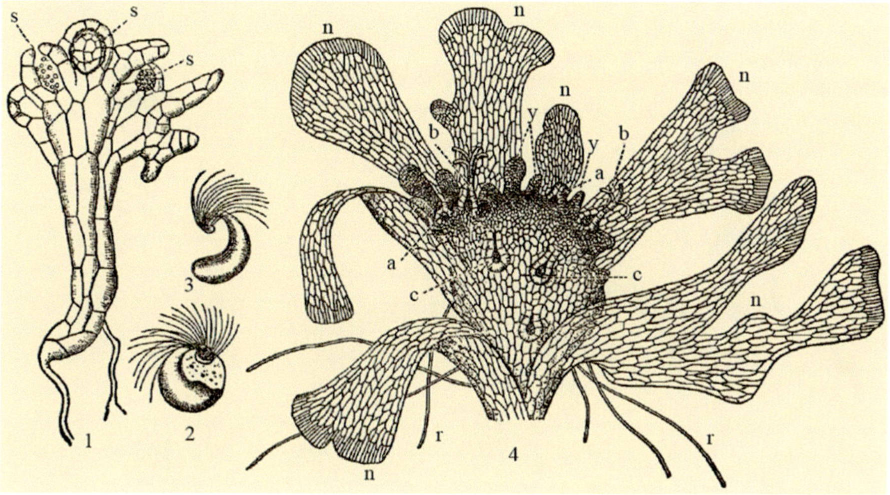

Fig. 6.28 The female and male gametophytes of *Equisetum arvense*. (1) A male gametophyte, s: antheridia. (2) A sperm that is still in the antheridium. (3) A mature sperm. (4) A female gametophyte—a: developing young archegonia, b: well-developed archegonia, c: unfertilized old archegonia, n: primary nutrient leaves, t: rhizoids, y: young prothalli. (After Figure 313 in Ikeno 1906)

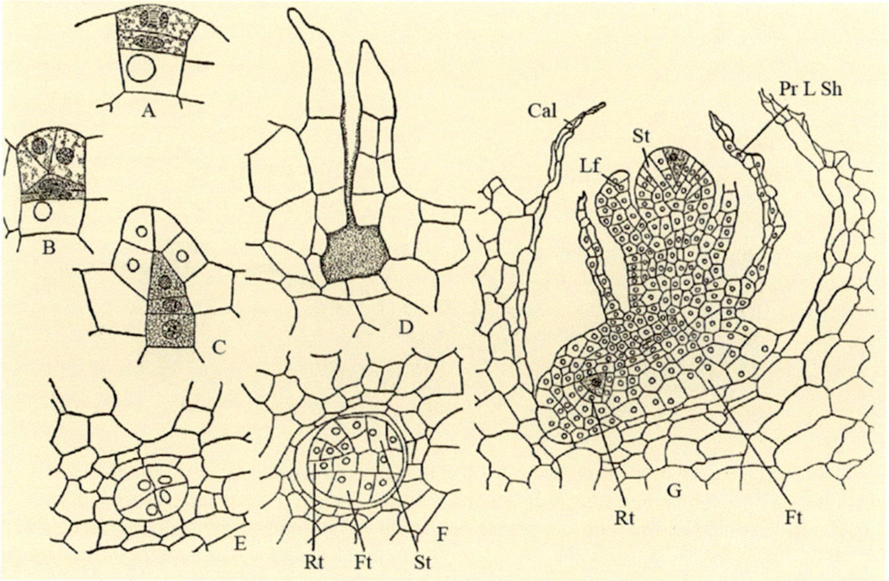

Fig. 6.29 Stages in embryo development on horsetail's prothallus. (**a–c**) *Equisetum telmateia* Ehrh. (**d–g**) *Equisetum debile* Roxb—(**d**) open archegonium, (**e**, **f**) young embryos, (**g**) embryo at an advanced stage of development (Cal: calyptra; Ft: foot; Lf: leaf; Pr, L, and Sh: primary leaf sheath; Rt: root; St: stem) (After Figure 140 in Smith 1938)

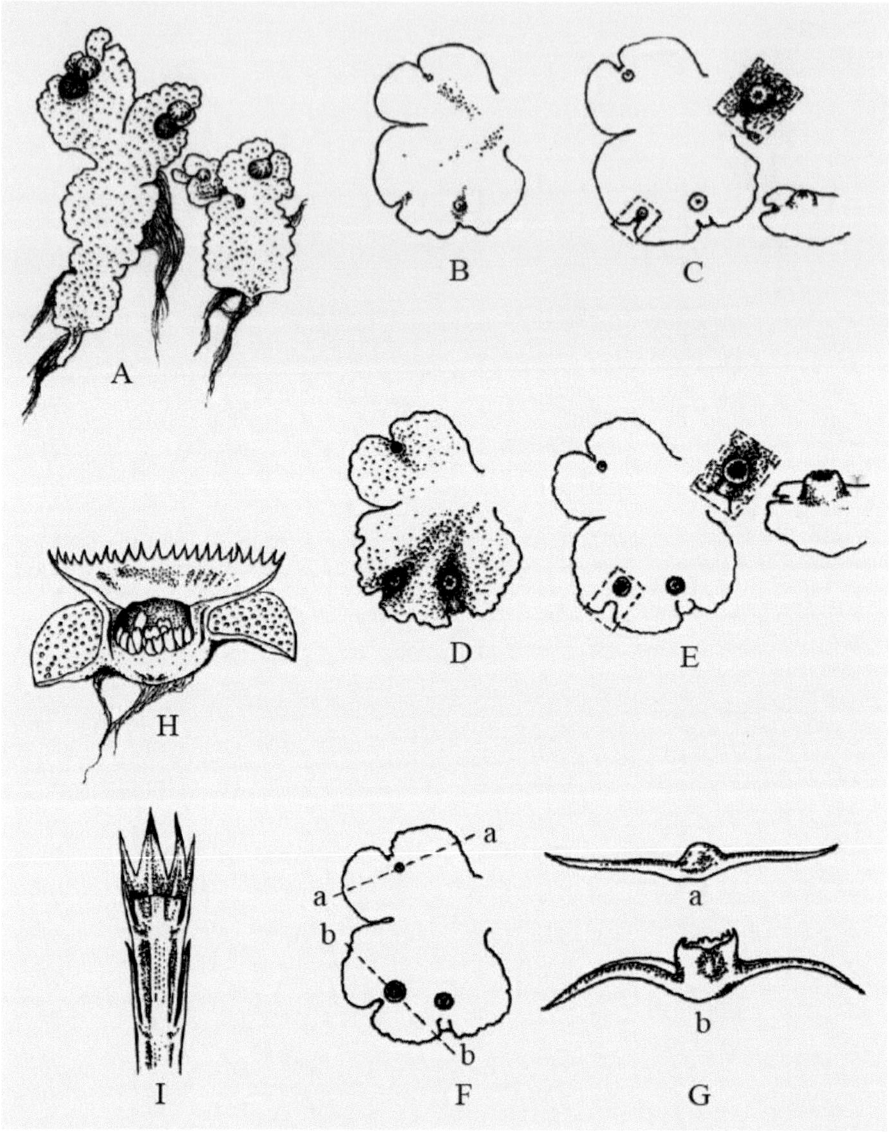

Fig. 6.30 A comparison of liverworts' bud dish with horsetails' secondary axes. (**a**) The position of bud dishes on *Marchantia polymorpha*; (**b**) a newly born bud dish; (**c**) an enlarged part of the newly born bud dish; (**d**) a bud dish with an opening on the top and a whorl of leaf clefts; (**e**) an enlarged part of the bud dish with an opening at the top and a whorl of leaf clefts; (**f**) a comparison between a juvenile and a mature bud dish; (**g**) same as in (**f**) but showing a transection; the inner cavity has formed; (**h**) a transection of a bud dish, showing asexual buds growing at the bottom of the inner cavity of the axilla; (**i**) a longitudinal section of a lycopod stem, showing the status of secondary axes similar to the bud dish of liverworts

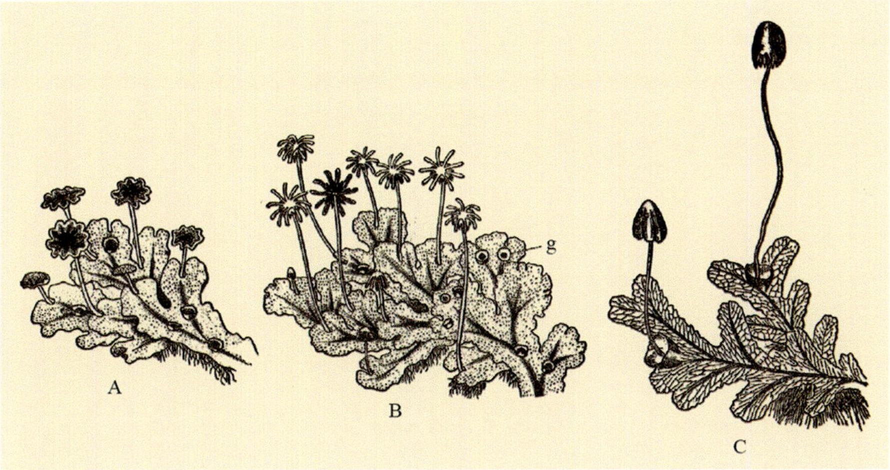

Fig. 6.31 *Marchantia polymorpha* and *Fegatella supradecomposita*. (**a**) Male gametophytes of *Marchantia polymorpha*, with an antheridial stalk; (**b**) female gametophytes of *Marchantia polymorpha*, with an umbrella-like archegonial stalk and a bud dish (g); (**c**) *Fegatella supradecomposita*'s female gametophytes and sporophytes in the archegonia with umbrella-like sporangiophore; fertilized eggs sprout to give sporophytes. (After Figures 243 and 245 in Ikeno 1906)

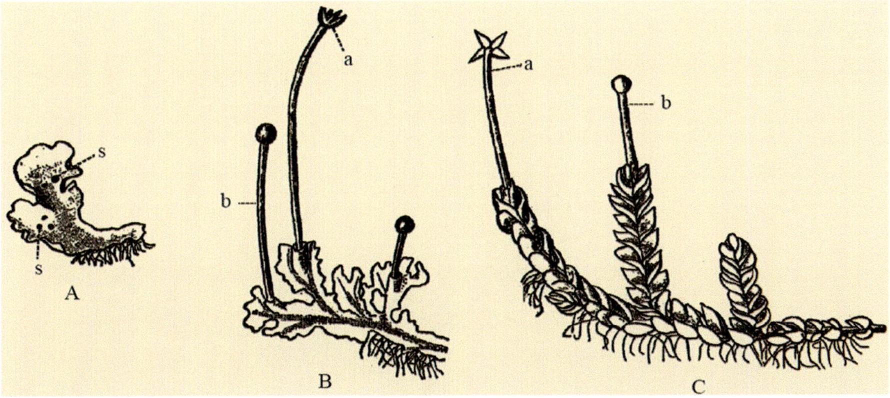

Fig. 6.32 *Pellia epiphylta* and *Nardia* sp. (**a**) *Pellia epiphylta*'s male gametophyte, s: antheridia; (**b**) *Pellia epiphylta*'s female gametophyte—a: sporophyte with an open capsule, b: sporophyte with a closed capsule; (**c**) *Nardia* sp., showing differentiated leaf clefts on the gametophytes—a: sporophyte with an open capsule; b: sporophyte with a closed capsule. (After Figure 238 in Ikeno 1906)

in the young embryo clearly demonstrate the relationship of the secondary axes with the splitting axes. The structure Smith called "primary leaf sheath" is the first leaf clefts from the first axis. The horsetail's young seedlings and the adult plants illustrated in Figs. 6.33 and 6.34 show more of the overlapped axes and whorls of

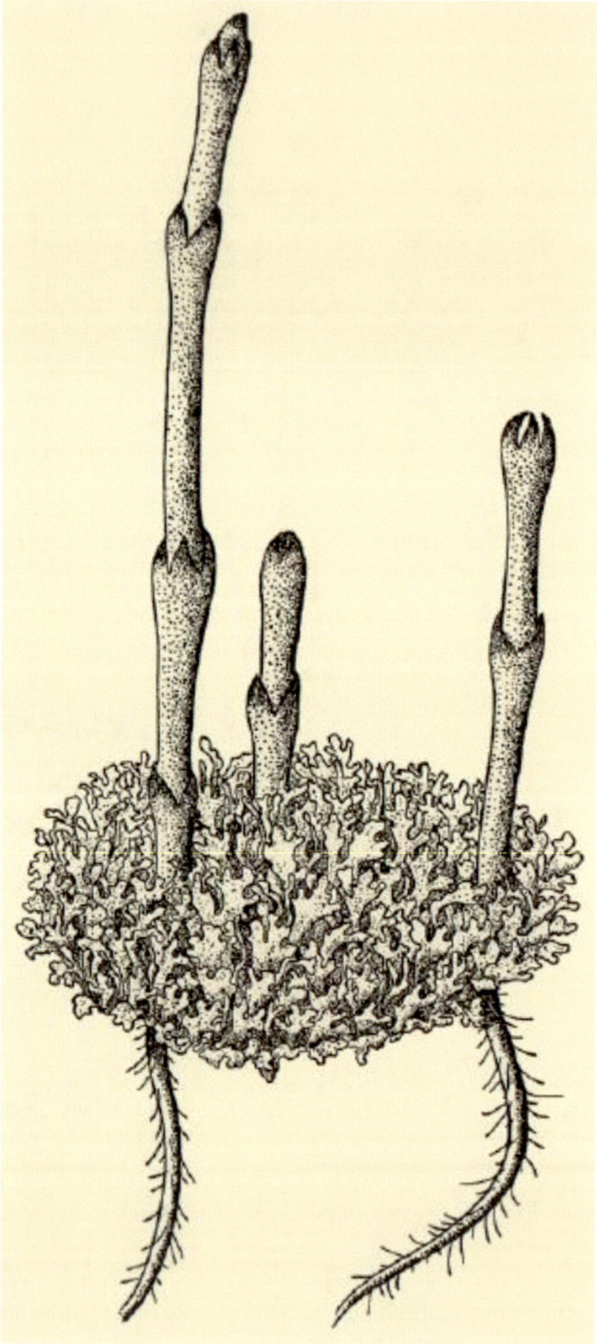

Fig. 6.33 Horsetails' gametophyte with young sporophytes. (After from Figure 161 in Smith 1955)

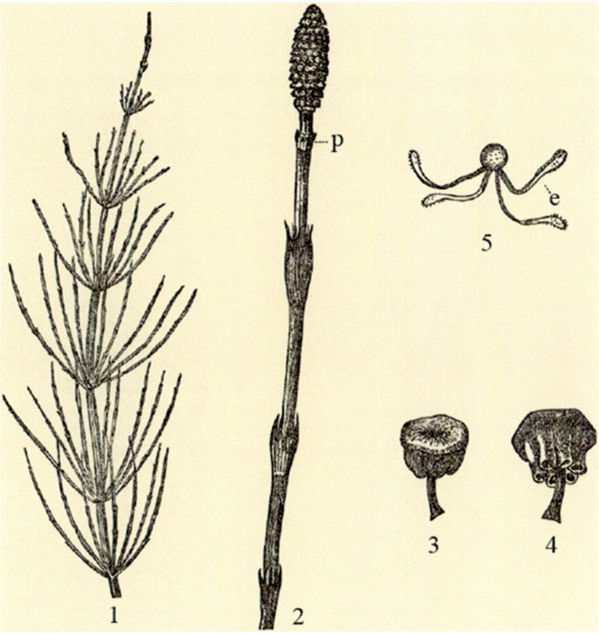

Fig. 6.34 *Equisetum arvense* L. (1) Vegetative shoot; (2) reproductive shoot, showing sporangial capsule and a whorl of leaf clefts on the top; (3) sporangiophore with dropping sporangia; (4) lateral view of the sporangiophore, showing six open sporangia; (5) sexual spore, e: elaters. (After Figure 311 in Ikeno 1906)

leaves at their nodes, a fact that the telome theory cannot explain. Therefore, Smith, a telome theory advocator, believed that the origin of horsetails' verticillate leaves was not clear. He also stated that Sphenophyllales and their relatives have cuneate leaves that do not have petioles but have dichotomous veins, that this kind of leaves appears to be a leaf-like branch, and that sterile single-vein leaves can be seen on many horsetails, which might be a degenerated sterile leaf-like branch. Using uncertain words such as "appear" and "might" indicates that his explanation is just speculation without any factual evidence.

Figure 6.35 shows a transection of a horsetail stem. We can see the detailed structures of vascular bundles, which are formed through secondary differentiation and extend from the base of the verticillate leaves into the internodes. This is not a uniformed stele, as the telome theory says, but multiple scattered tissues that are formed from the vascular bundles that extend into the internodes from each of the verticillate leaves, forming a ring of vascular bundle groups. This is also a negation of the telome theory.

Although many ferns evolved into giant trees long ago in the Mesozoic period, modern ferns, except for *Alsophila spinulosa* (Wallich ex Hooker) R.M. Tryon and a few other species of the same genus, are all herbs. Almost all have their multiple axes united at the lower portions to form a secondary stem, creep on the ground, and grow secondary roots downward. Even if they have an upright stem, it has undeveloped cambium, becoming a short herb. Only the scaly tree ferns stand vertically on the ground, forming thick small trees and growing fibrous roots downward around the base. The morphogenic difference between the two kinds of stems is due to their

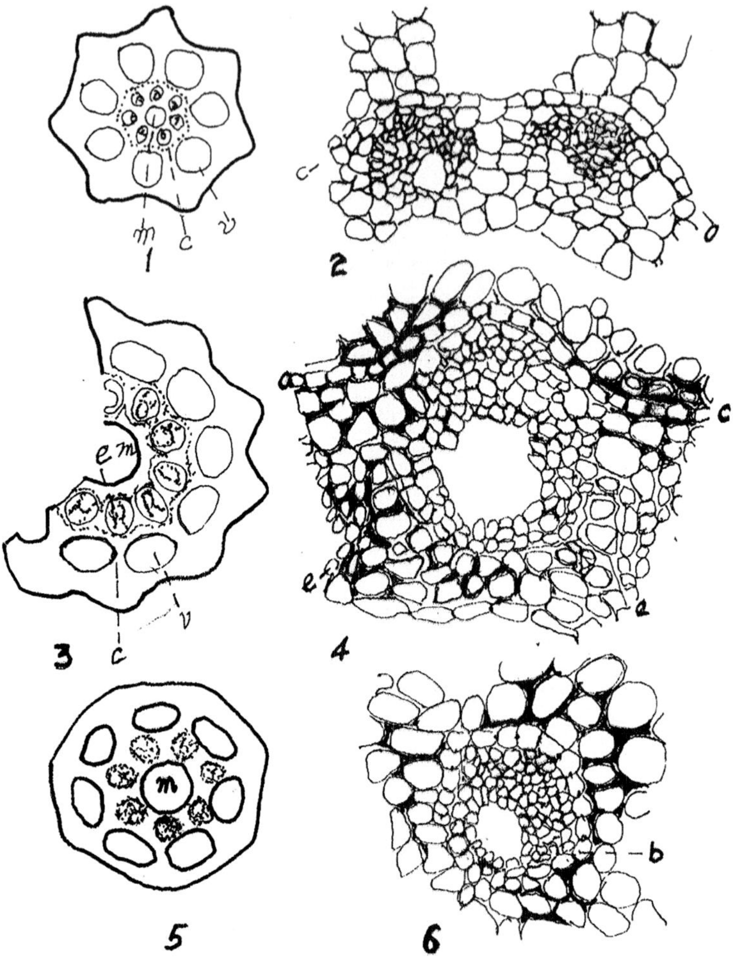

Fig. 6.35 The transection of the stems in the genus *Equisetum* (1), (2) *Equisetum papusire*, (3, 4) *Equisetum sylvaticum*, (5, 6) *Equisetum litorole.* (m) Medullary cavity, (v) vallecular cavity, (c) endodermis outside vascular bundles; (e) same as (c), indicating the inner part of endodermis; (b) endodermis surrounding vascular bundles. Scale: (2, 5), 26/1; (6), 160/1

slight difference in the growing pace of the so-called intercalary meristem during the formation of the stems via the union of the lower portions of their axes. In the scaly tree ferns, before forming the vascular bundles via secondary differentiation, cells in the united lower intercalary meristem divide vigorously to form a large truck. Every axis follows the same differentiation model to form a large upright pole-like stem, which makes the scaly tree ferns become short trees. In the herb ferns, the cells in the intercalary meristem do not divide so rapidly; their vascular bundles differentiate earlier. The growth of the meristem in the lower portion of the upper axis only causes the stem to extend and become an internode. Most of them do not develop cambium, do not have secondary thickening, and thus become slimmer. In some lycopods and ferns that have cambia and thus can make metaxylem, the division of the cambia is very weak and ages quickly. Their stems no longer thicken. Most of them cannot stand upright but creep on the ground.

References

Ikeno S (1906) Plant systematics (Chinese translation by Luo Zongluo published in 1947 published by The Commercial Press, Shanghai)
Smith GM (1938) Cryptogamic botany, Algae and fungi, vol I. McGraw-Hill, New York
Smith GM (1955) Cryptogamic botany Vol. II. Bryophytes and pteridophytes, 2nd edn. McGraw-Hill, York and London

Chapter 7
The Ontogenesis of Higher Plants: Spermatophyte

Plants have evolved into Spermatophyte (higher seed plants), the gametophytes of which are completely parasitic in the spores. This is a reversal of what has happened to bryophytes. A prothallus stage, with green, independent gametophytes, does not exist. The parasitic gametophyte on the sporophyte, after the egg is fertilized by a sperm, develops into the sporophyte of a new generation. The first axis of the new gametophyte forms the first leaf clefts, the specialized cotyledons, on the apex, and the lower end of the axis evolves into roots or degenerates to the cuff of the secondary roots. Another cell that grows with the egg is fertilized by another sperm and then develops into endosperm, the nutrient source for the growth and development of the young embryo. The event of simultaneous fertilization of the egg and the binuclear embryonary sac cell in the parasitic sporophyte on the gametophyte is called double fertilization, which is unique to the seed plants. The female gametophyte and the dormant seedling and endosperm that have developed from the fertilized female gametophyte all parasitize on the sporophyte of the last generation. They are wrapped by leaf clefts of several axes, forming a dormant reproductive organ, the seed. Depending on whether their seeds are bare or wrapped in the ovary, the seed plants are classified into two broad categories: gymnosperms and angiosperms. Let us take a look at how the primary and secondary axes of gymnosperms and angiosperms build their organs.

7.1 Gymnosperm

Most of the modern gymnosperms (plants with naked seeds) are plants with very strong secondary vascular bundle systems; the vast majority of them are tall trees. Because of the strong development of the secondary vascular bundles, they have masked the primary tissue structure. As a result, their segmental stem is regarded as a unified whole. Their leaf clefts, except for the cycad family, no longer develop into a structure of leaflets by secondary growth.

C. Yen, *Constitution of Organs of the Higher Plants*,
https://doi.org/10.1007/978-981-19-1685-4_7

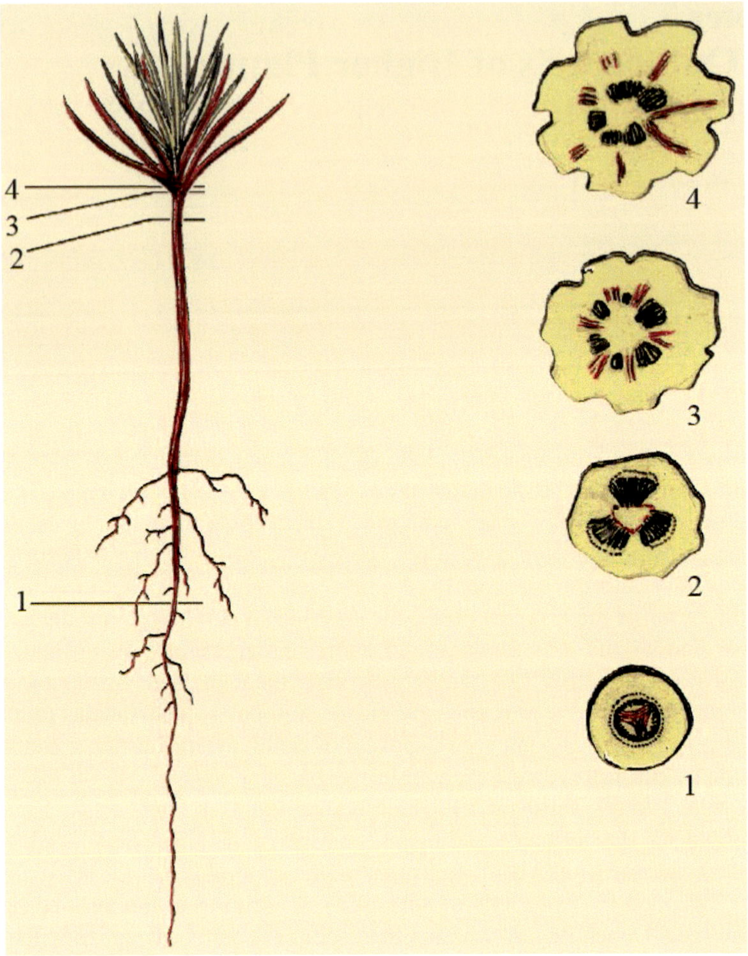

Fig. 7.1 The seedling of *Pinus massoniana* Lamb., showing multilobed cotyledons and the secondary xylem and phloem, which are well developed at an early stage, characteristics that make pine a tree. (Colors in the figure are not natural; they are added there only to differentiate the vascular bundles of different axes)

Pinus is a group of popular gymnosperms. Its first axis develops into a radicle at the lower end and multilobed cotyledons at the top (Fig. 7.1). At the early stage of germination, the cotyledons are also the absorption organ that absorbs nutrients from the endosperm. After the formation of the seedling, the main root (the radicle), together with multiple secondary roots developed from the stele of the radicle, forms a strong root system to absorb water and inorganic nutrients from the soil. The cotyledons extend out of the ground and become the first whorl of green cotyledons, which synthesize organic matters by photosynthesis. Leaf clefts above the second axis become alternative leaf clefts due to their unbalanced growth and twisted rise.

Fig. 7.2 The clustered branches of *Pinus radiate* D. Don. (**a**) A pine branch of a 3-year-old: pine needles in the axillae have fallen off, and only the scaly leaves remain in the first section; needle leaves grown in the axilla of the scaly leaves in the second section; the axillary growth of needle leaves has not started, and only scaly leaves are visible, wrapping the stem axis. (**b**) A scaly leaf. (**c**) Five levels of secondary axes grown from the axillae of the scaly leaves; the lower four levels split to form opposite scaly leaves, and the fifth splits to form three needles

These leaf clefts become simple microphylls. Twigs develop from buds at higher leaf axillae. These twigs are actually leaf clusters formed from multiple axes by splitting (Fig. 7.2c). The top axis, depending upon the species, becomes di-, tri-, or pentachotomous and elongates to become green, photosynthetic "pine needles." The leaf clefts at the lower axes become scaly leaves. These vegetative twigs basically do not form a stem and are completely detached as a whole after aging. This is the typical structure of Pinaceae.

Leaf clefts on the main stem or branches become scaly leaves (Fig. 7.2b); the xylem and phloem of the secondary vascular bundles grow and develop rapidly to form the woody trunks and the thick cortex. In the next growing period, while a new stem grows on the top, the aforementioned pine needles will grow from the axillae of the scaly leaves, too. This regularly regulated physiological mechanism should also be the mechanism that makes pine a tall tree. The genes that regulate the physiological mechanism should be a part of the main genes that make pine a tree. As we all know, plants should undergo a physiological transition when enterring the blooming stage, such as vernalization in wheat. This is also another important physiological process that makes pine a tree. For pine, it takes several years to more than 10 years to complete this process, unlike wheat, which completes vernalization after one low thermophase. This is also where tall trees differ from annual or biannual herbs. The cambium of vascular bundles in a tree trunk continuously spawns new xylem inwardly and phloem outwardly, making the trunk thicker and larger. This secondary growth can last for hundreds or even thousands of years, making pine a perennial tall tree. Sequoia of America, which is similar to pine, can not only grow for thousands of years but also be more than 200 m in height with a trunk of 3–4 m in diameter.

Similar to pines, cypress and fir are also trees. Their morphology and physiological structure are also similar to pines. However, they do not have scaly leaves and needles, which make them apart from pines. Their green leaves are microphylls with simple single veins and without secondary axial structures above the second level. However, as mentioned above, their macrosporophylls are compound leaves composed of multiple levels of secondary axes (Fig. 7.3a–f, m); their vegetative green leaves are oppositely dichotomous (Figs. 7.4 and 7.5) or are polychotomous verticillate microphylls (Fig. 7.3g). Organ structures like these are completely inconsistent with the concept of the telome theory but clearly show the pattern of multiple secondary axes.

Ginkgo is an ancient gymnosperm. It flourished in the Mesozoic era. However, there is only one species, *Ginkgo biloba*, left today. It has survived the cold in the ice age in southwestern China. Nowadays, it has been planted all over the world, indicating its strong adaptability. Ginkgo has dichotomous leaf veins (Fig. 7.7a), which seems to fit the telome theory. However, it should not be forgotten that modern physiological experiments have proved that leaf veins are secondary structures and are not formed from branches, and it is also of the normal state that the dichotomous branching is formed by the merging of two auxin tracts. From its phyllotaxy, one can see that its cotyledons are oppositely dichotomous (Fig. 7.6); the opposite scaly leaves of its axillary buds are clearly visible (indicated by the arrowhead in Fig. 7.7); and its alternate phyllotaxy is composed of opposite leaves formed by unequal growth and dislocation. First, there is opposition then there is alternation, which is antagonistic to the viewpoint of the telome theory that the opposite results from the convergent aggregation of the alternate. Viewed from the petiole's transection (Fig. 7.7), the leaf veins show the shape of the split axes. Ginkgo shows that its organs are composed of multiple secondary axes. The opposite

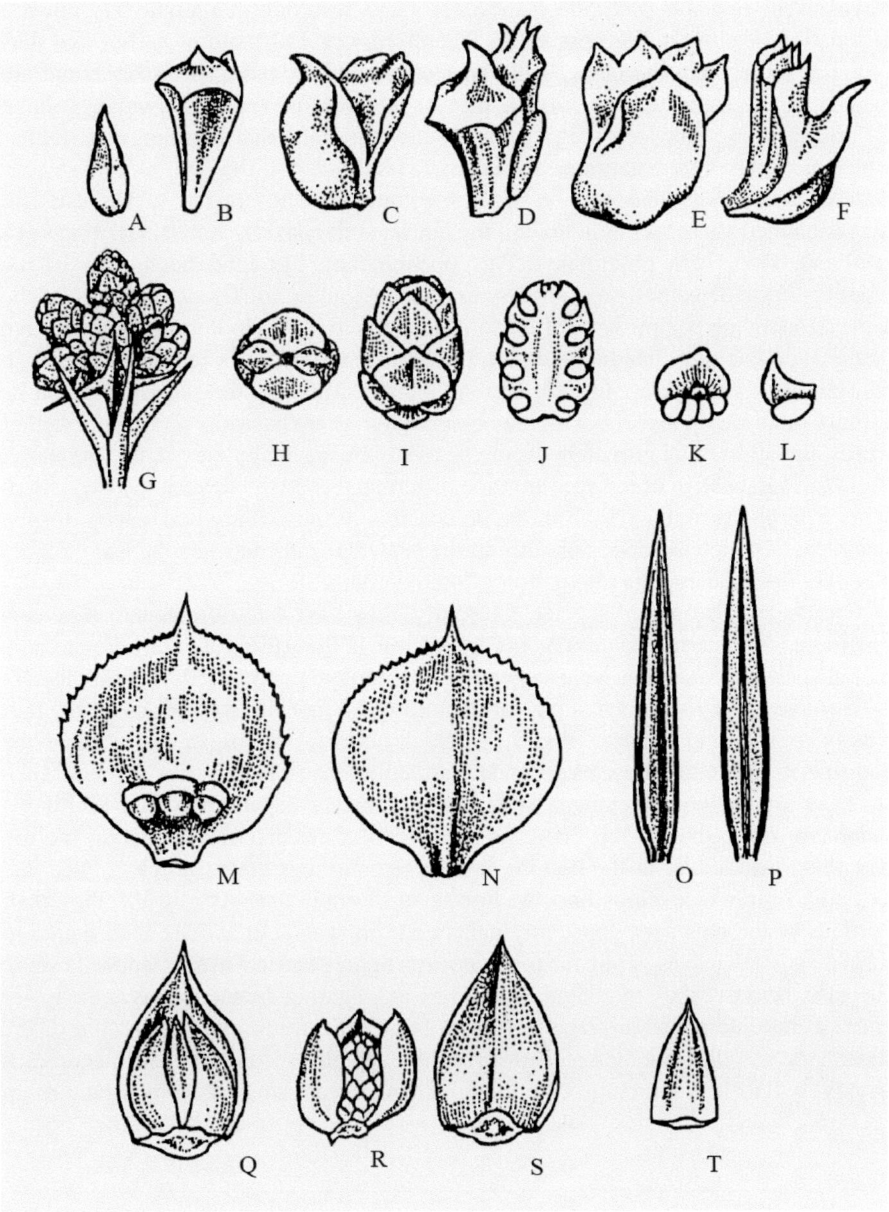

Fig. 7.3 *Cryptomeria fortunei* Hooibrenk ex Otto et Dietr. and *Cunninghamia lanceolata* (Lamb.) Hook. (**a–l**) *Cryptomeria* fortune—(**a**) a sterile macrosporophyll developed from a simple leaf cleft; (**b**) a macrosporophyll developing by adaxial fusion of two levels of leaf clefts; (**c**) a macrosporophyll developed through adaxial-abaxial fusion of three (1 + 2) leaf clefts of two levels; (**d**) a macrosporophyll developed through adaxial-abaxial fusion of four (1 + 3) leaf clefts of two levels; (**e**) the ventral view of a macrosporophyll developed through the adaxial-abaxial fusion of five (1 + 4) leaf clefts of two levels; (**f**) same as (**e**) but from the side view; (**g**) a male fruit branch, showing three verticillate needle leaves; (**h**) the bottom view of a male fruit, showing opposite microsporophylls; (**i**) same as (**h**) but from the side view; (**j**) the longitudinal section of (**i**); (**k**) a

leaves in the opposite phyllotaxy unevenly grow, resulting in alternatively upward dislocation, leading to the opposition. Therefore, one leaf matures earlier and thus becomes larger than the other. The leaf clefts at the second axis also rise and are dislocated the same way to form a pair of alternate leaves. Meanwhile, volume difference occurs between the portions of the leaf clefts that are merged together, which also causes the rotation of the alternate phyllotaxy. If the right secondary axis matures first, the right side will be larger in volume than the left side, which results in a right-handed alternate phyllotaxy; if the left secondary axis matures first, it makes a left-handed alternate phyllotaxy. This phenomenon has long been seen in the organogenesis of some lycopodiophytes. As for ginkgo, different branches of the same tree can have a left-handed or right-handed alternate phyllotaxy. It seems that being a left- or right-handed alternate phyllotaxy in ginkgo is random rather than constantly genetically controlled. Ginkgo's cotyledons and the scaly leaves on its axillary buds are opposite, but all the other euphyllae are basically alternate. In other words, the differential growth between the two opposite euphyllae occurs very early. From the perspective of the seedling, the differential growth between the two leaves starts from the second axis. From the perspective of the axillary bud, except for the scaly leaves which are opposite, differential growth occurs between the leaf clefts of all euphyllae, causing the alternation of the euphyllae.

Cycads are also extant species. Currently, they have only two genera. *Cycas* is distributed in eastern Asia, and *Zamia* is endemic to the Americas. They are the only gymnosperm species that have vegetative macrophylls composed of multiple secondary axes. They still have a pair of dichotomous opposite cotyledons on the first axis of their young embryo (Fig. 7.8), which becomes the organ that absorbs the endosperm. No matter how its secondary structure conceals its native condition, the structure of the young embryo still clearly shows the pattern of the multilevel secondary axes; that is, the first axis splits to become two cotyledons on the top and develops radicles at the bottom, and the second axis grows in the cotyledons' axilla and further develops into the first level of euphyllae. As a result, the upper portions of the new axes constantly generated this way split and are then flatted to form new green leaves, while the lower portion aggregates to form a composite stem. Because the intercalary meristems formed by aggregating the meristems at the lower ends of multiple axes converge to form a large stem trunk, but the cambia of the secondary vascular bundles stop to grow thicker, the secondary stem becomes a woody cylindrical structure. We have already seen such a stem structure in

---

Fig. 7.3 (continued) dorsal view of a microsporophyll, showing four microsporangia; (**l**) same as (**k**) but from the side view; (**m–t**) *Cunninghamia lanceolata*; (**m**) a dorsal view of a microsporophyll, showing three macrospores (seeds) grown on three parallelly arranged, adaxially-abaxially fused secondary leaf clefts; (**n**) same as (**m**) but from a ventral view; (**o**) the dorsal view of a nutrient needle leaf; (**p**) the dorsal view of a nutrient needle leaf; (**q**) a microsporophyll and an axillarilly grown microsporangium wrapped by scaly leaves; (**r**) same as (**q**) but showing the microsporangium; (**s**) same as (**q**) but with a dorsal view of the microsporophyll, showing that the microsporophyll is not fused with the microsporangium; (**t**) a sterile microsporophyll

Fig. 7.4 The seedling of *Cupressus duclouxiana* Hickel, showing the dichotomous cotyledons and the vascular bundle system. (It clearly shows a dichotomous secondary axis pattern; the red color highlights the first axis, the vascular bundles of the cotyledon axis, and the strong arborous development of secondary vascular bundles. Colors in the figures are not natural; they are added there only to differentiate the vascular bundles of different axes)

Cyatheaceae. The radicle and secondary roots develop into a lignified stout fibrous root system. The multiple secondary axes develop into pinnate compound leaves. Such a calendrical stem and the multiaxial pinnate compound macrophylls constitute the special morphological characteristics of cycads. The morphological structure of cycads is unique among gymnosperms. However, their stems and leaves are similar to the morphological structure of the palm family in angiosperms. The vascular bundles of the leaf clefts extend downward into the stem. However, the xylem and

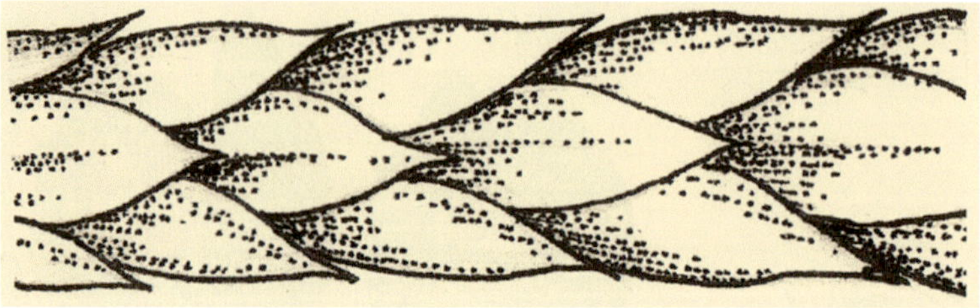

Fig. 7.5 A portion of a small branch of *Cupressus duclouxiana* Hickel, showing opposite phyllotaxy

phloem do not form cambia for continuous growth, and thus the stem cannot continue to grow thicker and larger. All these are completely different from the stems of pine and ginkgo.

Cycas' sporophytes are dioecious. As shown in Fig. 7.9b, d, like the multiaxis pinnate vegetative green leaves, the macrosporophyll and microsporophyll are also composed of multiple secondary axes. However, they are very different from the former in morphology. A large number of microsporophylls aggregate at the apex of the male plant, forming a large microstrobilus (Fig. 7.9a). The macrosporophylls also aggregate at the apex of the female plant (Fig. 7.10). While they are similar to the vegetative green leaves in morphology, unlike the macrosporophylls, the macrosporophylls do not aggregate into a ball. This is also the big difference between the *Cycas* of Asia and the *Zamia* of the Americas. On the dorsal side of *Cycas*' microsporophyll, dense secondary axes develop into microsporangia (Fig. 7.9c). The microspore mother cells formed in the microsporangia undergo meiosis to produce microspores, i.e., the male gametophytes, which completely parasitize on the sporophyte of the previous generation. On the macrosporophylls near the two sides of the petiole, there are seed coats evolved from three to four groups of multiple secondary axes. In the center, the egg mother cell formed from a secondary axis undergoes meiosis to produce an egg, which, together with the synergids and other embryo sac cells in the embryonic sac, form the gametophyte. The macrogametophyte also parasitizes within the seed coat of the sporophyte of the previous generation. A sporophyte, i.e. the young embryo and endosperm, is formed after double fertilization. The sporophyte of the new generation, together with the seed coat of the previous generation, develops into a dormant reproductive organ, the seed.

Ephedra is a shrub in gymnosperms. It has the typical structure of the multiple secondary axis pattern, composed of opposite leaf clefts interconnected one after another (Fig. 7.11). Although its vascular bundle system is also well developed with a high degree of lignification, the cambium does not continue to differentiate xylem and phloem, and thus, its stem cannot continue to thicken. As a result, it is a short shrub. Its stem completely preserves the initial morphology of the secondary axis pattern, which is not buried by the secondary structures.

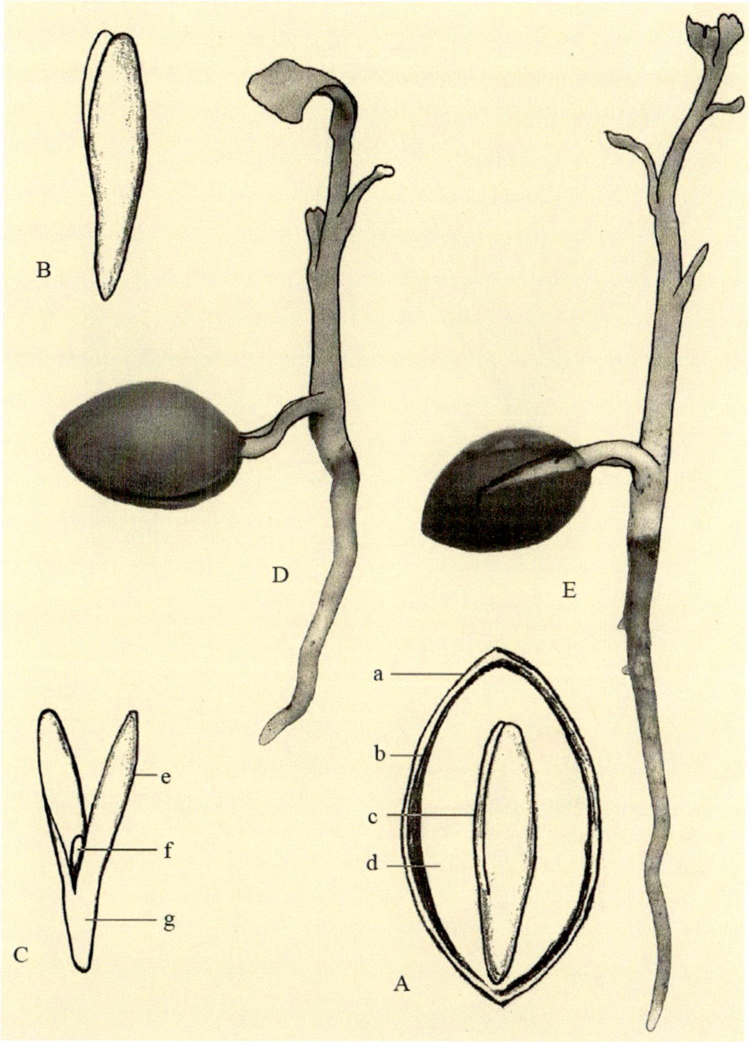

Fig. 7.6 *Ginkgo biloba* L., showing a young embryo buried within the endosperm and seedlings that have just emerged from the soil. (**a**) The seed; a: the outer seed coat, b: the inner seed coat, c: the dichotomous opposite cotyledons formed on the upper portion of the first axis of the embryo and hypocotyl and radicle formed from the unsplit lower portion. (**b**) A young embryo separated from a seed. (**c**) A young embryo separated from a seed with cotyledons (e), which is artificially opened to show the euphylla (f) formed from the second axis and the radicle (g) formed from the lower portion of the first axis. (**d**, **e**) Young seedlings, showing the dichotomous leaf clefts on the axes above the second axis. A pair of unevenly developed young leaf clefts continue to grow and develop, becoming the opposite by rising and displacing

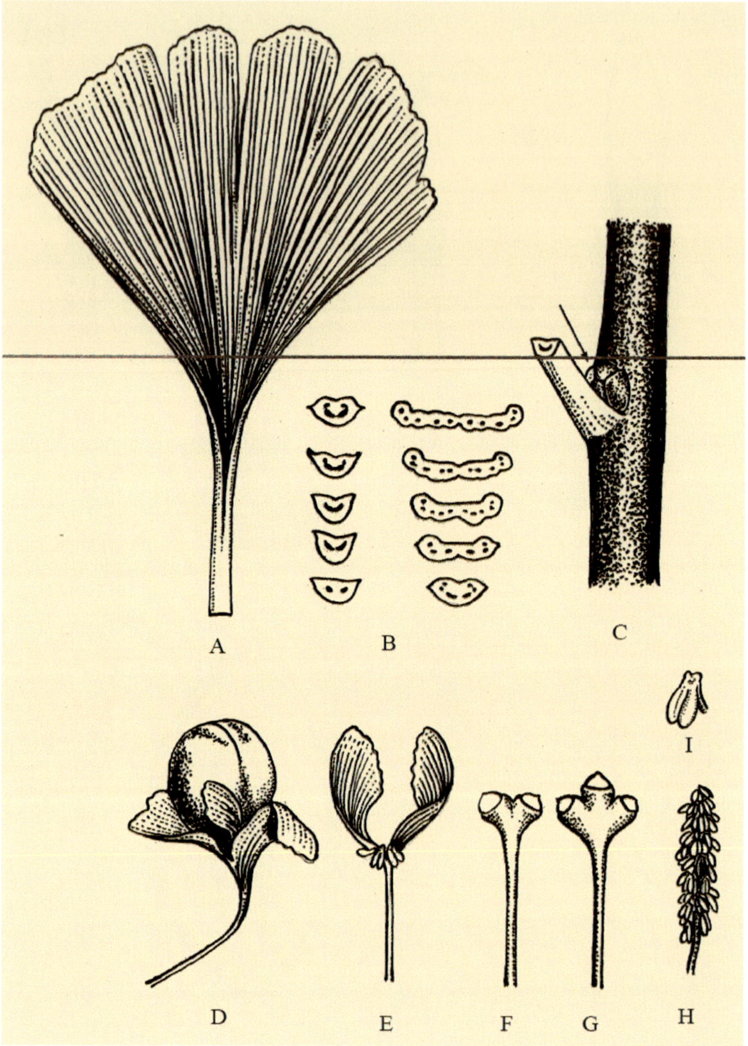

Fig. 7.7 The leaf organs of *Ginkgo biloba* L. (**a**) A leaf, showing dichotomous branch veins; (**b**) a transection view of the petiole, showing that dichotomous branch veins converge into one bundle, which then diverges again into two bundles growing into the stem to unite with other vascular bundles there; (**c**) an axillary bud in the leaf axilla; the arrowhead points to the opposite scaly leaves; (**d**) a seed grown on a leaf; (**e**) a microsporangium on a leaf; (**f**) a microsporophyll with a pair of ovules; (**g**) a microsporophyll with a pair of ovules and the third ovule; (**h**) a microsporangium order; (**i**) a microsporangium. (**a–d** are after Figure 383 in Ikeno 1906)

Welwitschia mirabilis Hook. f. is a peculiar gymnosperm plant endemic to the arid region of West Africa. It has a growth period of up to 100 years, so it is called "Orchid of 100 Years Old." Above its bilobed cotyledons (the leaf clefts on the first axis), the second axis splits to form two giant green leaves. The meristem tissues in

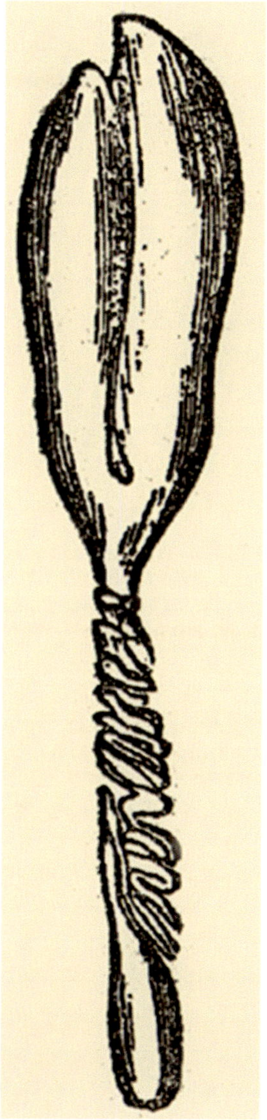

Fig. 7.8 A young embryo of *Cycas revolute* L., showing the dichotomous cotyledons and spiral crimped suspensor. (After Figure 368 in Ikeno 1906)

the lower portion of the second axis maintain meristematic growth, which makes the two green leaves constantly elongate and renew (Fig. 7.12). Besides plants with two leaf clefts, there also exist plants with multiple leaf clefts. The axillary axis developed in the axilla between the two green leaves (the secondary axilla) forms a reproductive branch (Fig. 7.13). The upper portion of the axis that forms the reproductive branch enlarges and becomes hollow but does not split completely, only showing semi-splitting, forming two opposite microphylls with merged lower portions. What *Welwitschia mirabilis* demonstrates is the completely typical

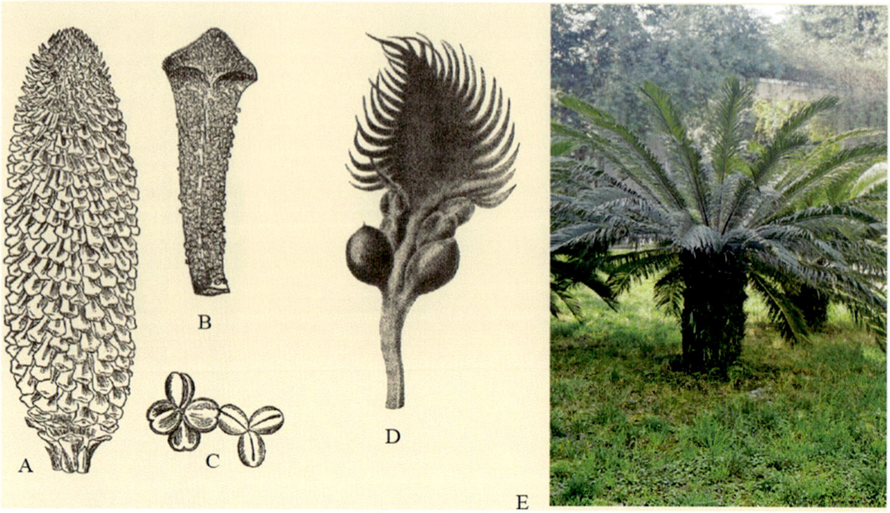

Fig. 7.9 The male and female sporophylls and the plant of *Cycas revolute* L. (**a**) A microsporophyll cone; (**b**) a microsporophyll; (**c**) a microsporangium; (**d**) a microsporophyll, showing two rows of six microspores on the leaf; (**e**) the whole plant. ((**a**–**d**) are after Figures 362 and 363 in Ikeno 1906)

Fig. 7.10 The female plant of *Zamia floridana* DC. and a microsporophyll with two microspores. (After Figure 358 in Ikeno 1906)

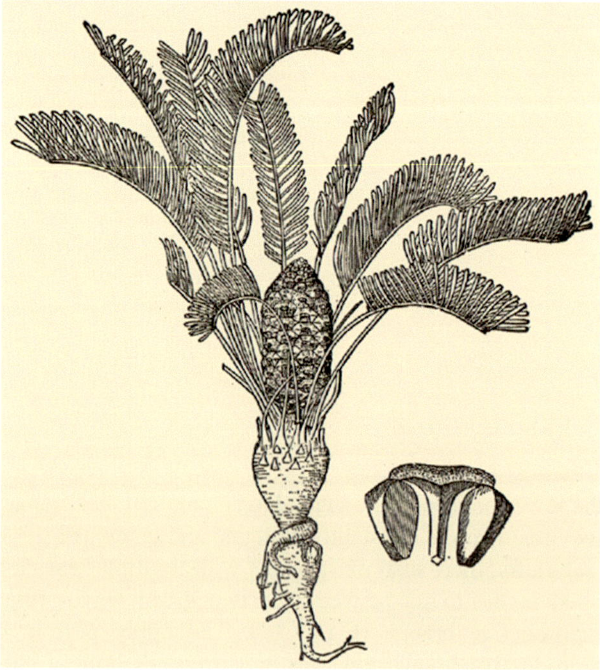

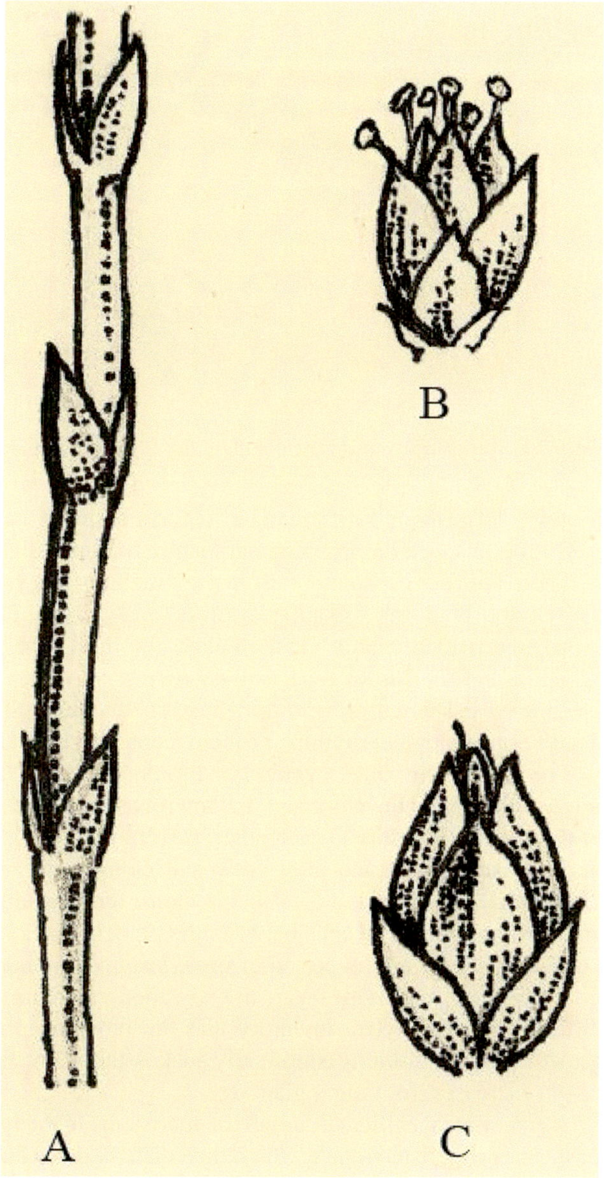

Fig. 7.11 *Ephedra intermedia* Schrenk ex Mey. (**a**) A portion of the stem, showing opposite leaf clefts; (**b**) microsporophyll corn, showing opposite microsporophylls; (**c**) microsporophyll corn, showing opposite macrosporophylls

secondary axial pattern of organogenesis. It is unique in that it consists of two huge green leaves composed of two single axes. Although the leaf tips are dead, the basal intercalary meristems continue to grow and regenerate, enabling the two huge green leaves to live for a hundred years. On the main root developed from the lower portion of the first axis, secondary roots grow. In the roots, the secondary xylems and phloem developed from the cambia formed by the secondary vascular bundles

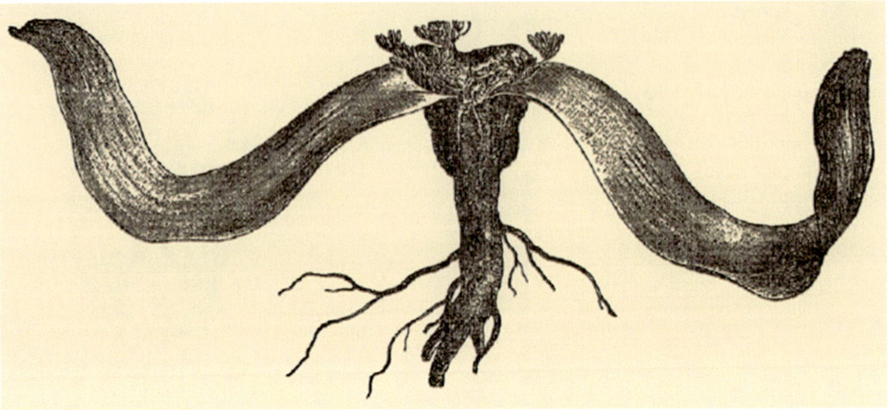

Fig. 7.12 *Welwitschia mirabilis* Hook. f. (After Figure 399 in Ikeno 1906)

continue to grow and differentiate to form a strong large main root system. This constitutes a special morphological feature unique to *Welwitschia mirabilis*.

There are four completely different structures in gymnosperms: the first type is the tall and thick trees of pine, cypress, and ginkgo; the second type is the small cycad tree, which cannot grow thicker; the third type is the herbaceous shrub of ephedra; and the fourth type is *Welwitschia mirabilis*, which has only two huge green leaves. The reason why they have such a huge difference in organogenesis is that the physiological rhythms of their secondary structural growth are different.

The reason why pine, cypress, and ginkgo become tall trees is that, at the early stage of their axial development, the interphase between the growth of the upper and lower axes is very close to each other; therefore, all the lower portions of the axes are in a meristem state, and they grow and elongate to form internodes. When they differentiate vascular bundles in the stem, the vascular bundles are alternatively arranged as a ring. The cells between the phloem and xylem of the vascular bundles keep dividing for a long period of time, forming cambial initial cells, which evolve into a circle of meristem layer, the so-called cambium. This cambium continually differentiates the xylem inwardly and the phloem outwardly. Such a continuous growth of the cambium continually makes the stem bigger and thicker while the height increases, forming a giant tree.

As mentioned earlier, although the meristems in the lower portion of the axes also form a large meristem mass, which then differentiates into a thick stem in *Cycas*, the stem does not have cambium and thus cannot grow thicker.

Although the structure of ephedra is similar to that of pine and cypress, its cambium is not as active as that of pine and cypress. Therefore, ephedra becomes a short herb-like shrub.

Welwitschia is very special. The second axis becomes two huge green leaves. The intercalary meristems in their lower portions continue to grow, making the two (or more) huge green leaves continue to elongate, which constitutes its unique morphology.

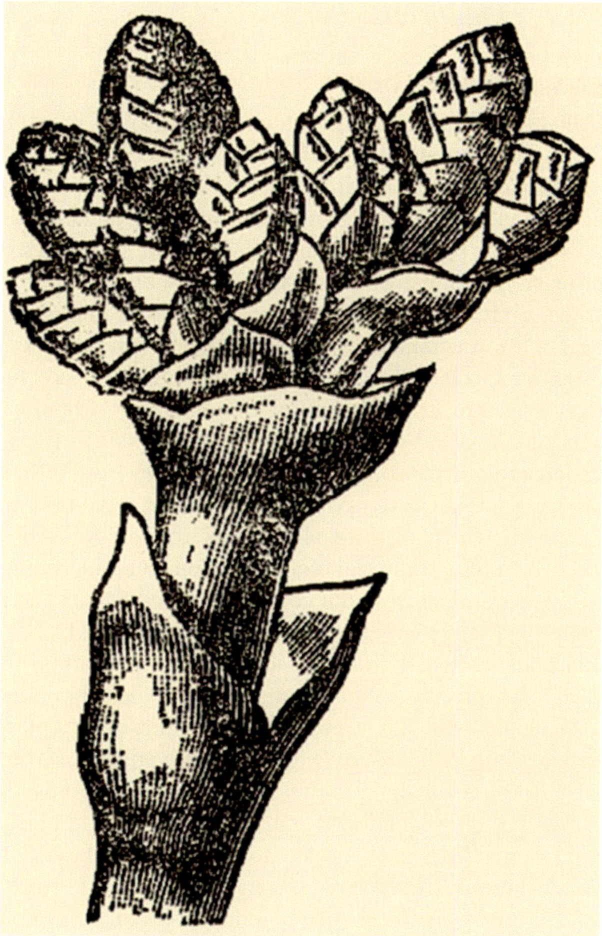

Fig. 7.13 A male branch of *Welwitschia mirabilis* Hook. f. and microsporophyll corns on it, showing opposite leaf clefts. (After Figure 400 in Ikeno 1906)

7.2 Angiosperm

There are two large groups of angiosperm plants, both forming cotyledons from the upper portion of the first axis. One group is simple, forming one cotyledon, i.e., the so-called monocots or monocotyledons. The other group is dichotomous, developing two cotyledons, i.e., the so-called dicots or dicotyledons. From the examination of a large number of monocotyledons and dicotyledons, we can see that not a single one of them follows the secondary axis pattern of organogenesis to form roots, stems, leaves, flowers, and fruits; none of them construct these organs following the telome theory.

7.2.1 Monocotyledons

For monocotyledons, let us discuss three representative types. The first is onion (*Allium cepa*) of Liliaceae, which is close to the prototype with little specialization. They are herbs with an underdeveloped secondary vascular bundle system, which does not cover up the primary structure. The second is palm (*Trachycarpus fortunei*) of Palmaceae, which is a small tree with a strong cylindrical secondary stem, which covers up its primary structure. However, its newborn seedlings still clearly show the primary structures, which are not covered by the secondary growth. The third one is wheat (*Triticum aestivum*) of Poaceae, whose immature embryo represents a higher degree of specialization. Comparing the immature embryos of these three very different species, one can see that, although they are very different in morphology, their axial structures are similar to each other. Although the upper portion of the first axis of onion has an absorption organ on its top for absorbing nutrients from the endosperm, this absorption organ largely develops into the first green leaf after being unearthed. The lower portion of the axis develops into a radicle, the organ for absorbing water and mineral nutrients from the soil. The newly developed secondary axis, the axilla of the leaf clefts, develops into the second green leaf, and the second root is formed outside the central core underneath the node (Fig. 7.14a). A young palm seedling is very similar to that of onion. Its first axis develops not into a green leaf after emerging from the soil but into a structure for absorbing endosperm. Its second axis is half-split to form the coleoptile. This coleoptile does not develop into a green leaf but has a protective function only. Only the third axis and beyond develop into green leaves (Fig. 7.14b). The first axis of wheat is fully specialized into an absorbing organ, the scutellum, dedicated to absorbing endosperm. The upper portion of the second axis is half-split to form a coleoptile, and a secondary root grows outside the stele on the node. The secondary root penetrates through the first axis, where the related tissue constitutes the coleorhiza (Fig. 7.14c).

The green leaf structure of onion (*Allium cepa* L.) (Fig. 7.15) of Liliaceae is basically similar to the aforementioned barley leaf and is also constructed from multiple secondary axes, but it is not as complicated as the green leaves of barley. The outer ligula has been degenerated, and the secondary axis that forms the leaf blade is no longer able to generate new axes, so there is no constriction on the leaf blade. The axis that forms the leaf blade is enlarged and hollowed but is not split, so it has a tubular structure (Fig. 7.15c, d). The leaves of garlic (*Allium sativum* L.) and garlic chives (*Allium tuberosum* Rottler ex Sprengel), which are in the same genus as onion, are only enlarged and flattened but not hollowed and split.

Let us look at palm of Palmaceae. The top of the first axis forms an absorbing leaf specialized for absorbing endosperm. The five to six vascular bundles secondarily differentiated in the upper portion transport the nutrients from the embryo down to the young seedling to support its growth and development. The base node is a compound node where multiple secondary axes converge. The lower portions of these axes are filled with meristems (Fig. 7.16). The base node formed by aggregation and the long-term growth of these intercalary meristems plays a key role in the

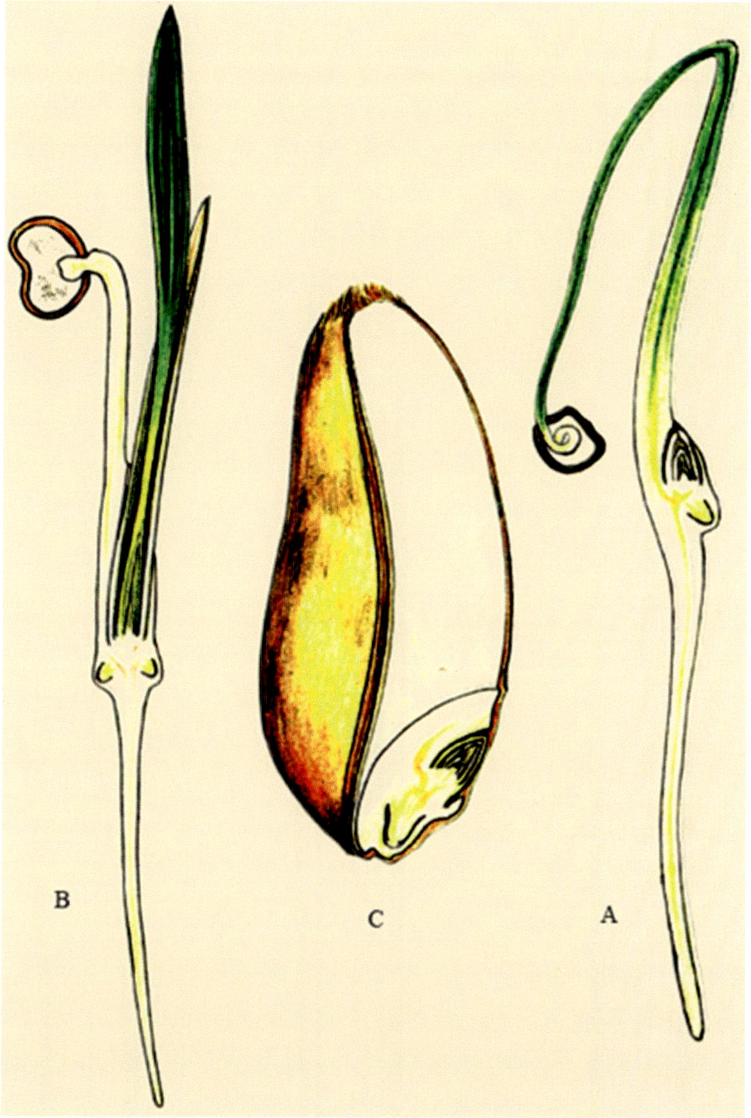

Fig. 7.14 Longitudinal sections of monocotyledonous embryos. (**a**) The first leaf cleft in the upper portion of the first axis of onion seedling becomes the first green leaf, its tip forms an endosperm-absorbing embryonic leaf, and the lower portion of the axis becomes the radicle; (**b**) the first leaf cleft in the upper portion of the first axis of palm seedling becomes the first green leaf, its tip forms an endosperm-absorbing embryonic leaf, and the lower portion of the axis becomes the radicle; (**c**) the longitudinal section of a wheat seed; it is clearly visible that the first axis as a whole becomes an endosperm-absorbing embryonic leaf, and all roots are secondary. Comparing with the longitudinal section of the onion seedling, a relatively consistent axial relationship is clearly visible

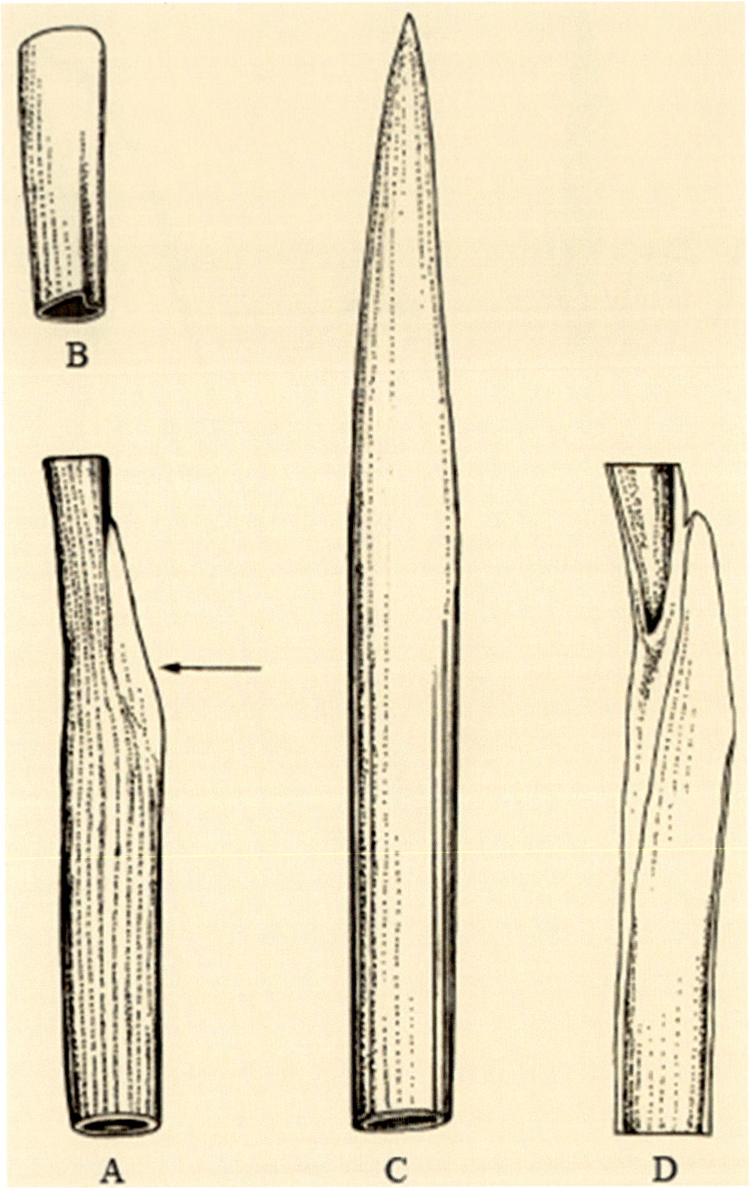

Fig. 7.15 The structures of onion leaves. (**a**) The upper portion of the sheath; (**b**) the lower portion of the enlarged hollow leaf blade; the arrowhead points to the inner ligula; (**c**) the middle portion of an enlarged hollow leaf; (**d**) the longitudinal section of the upper portion of the sheath and the lower portion of the enlarged hollow leaf, showing the relationships between the sheath, the blade, and the inner ligula. The outer ligula has been degenerated

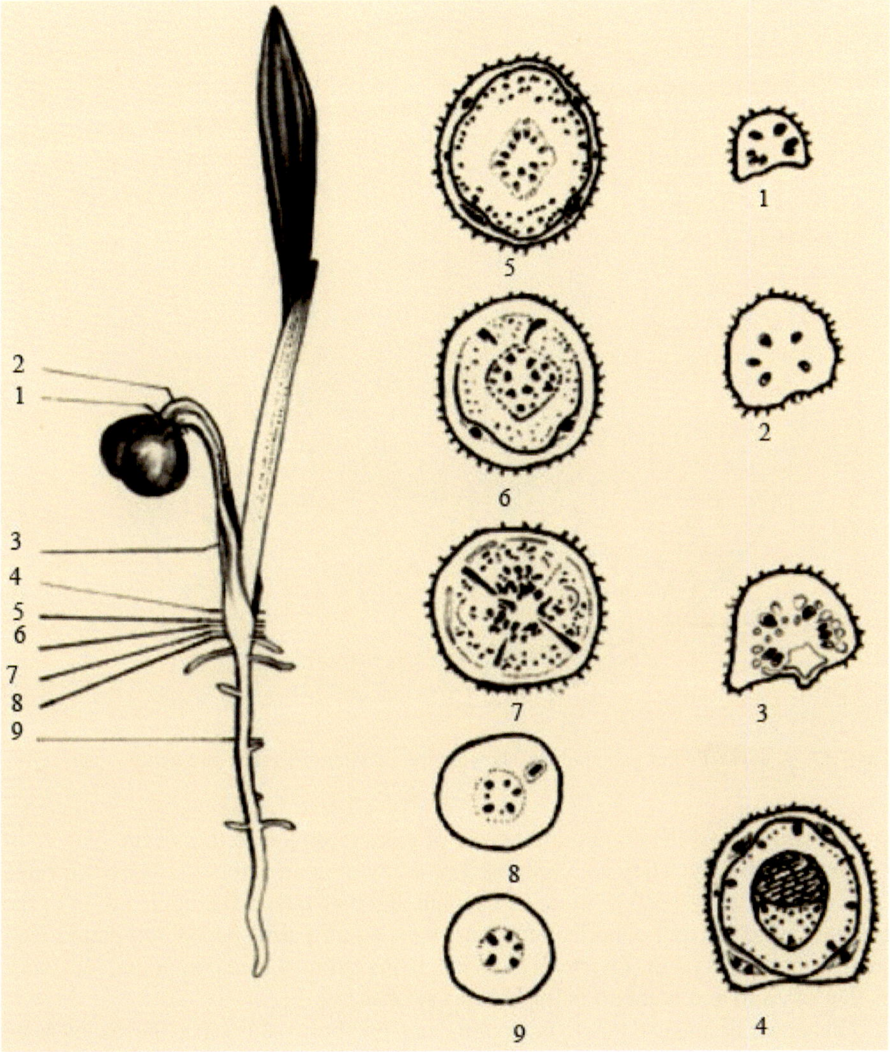

Fig. 7.16 A palm seedling and its transections. (The numbers on the figure only indicate the corresponding parts)

formation of strong thick stems. The secondary growth of these aggregated intercalary meristems thickens the multiaxial stem, making it a tree rather than an herb and also making it different from a dicotyledon tree in that it does not have cambium in its vascular bundles and thus cannot continuously thicken. After their formation, intercalary meristems differentiate into vascular bundles and the highly lignified pith cells between them, and thickening stops (Fig. 7.17). This intercalary growth is also unexplainable by the telome theory, but it is the exact proof of the objective existence of bipolarly growing secondary axes.

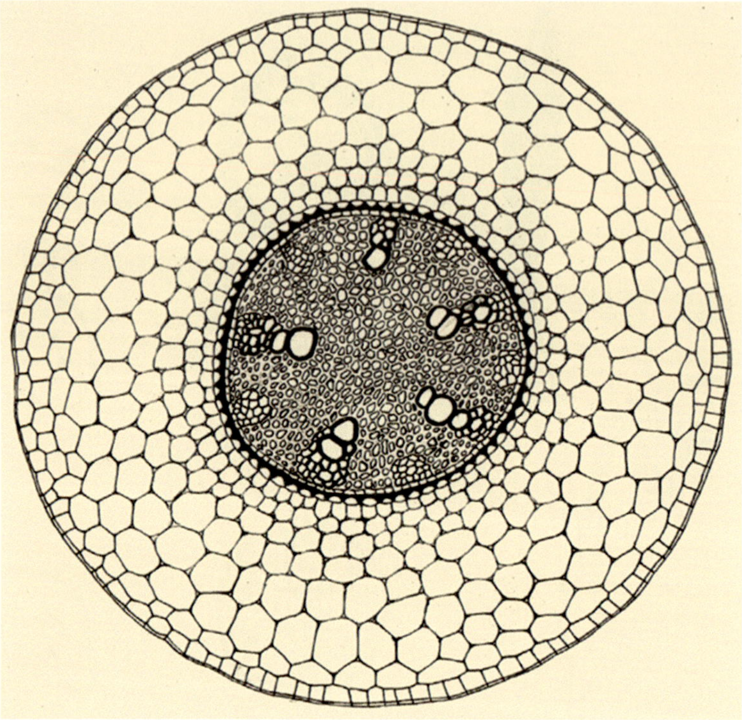

Fig. 7.17 A transection of palm radicle, showing the cambium-less stele and lignified pith cells

The structure of the compound leaves of palm is similar to that of the glumes of hooded-awn barley. They are composed of many pinnate compound leaves formed from multiple axes, such as those of coconut and date palm. The aggregation of the nodes of pinnate compound leaflets constitutes a half-palm-shaped compound leaf, such as that of fan palm (*Livistona chinensis*), or a palm-shaped compound leaf, such as that of palm and bamboo palm (*Rhapis excelsa*).

The inner ligulas of palm, fan palm, and bamboo palm have the exact same structures as those of Gramineae. However, further research is needed to tell whether the webbed protrusions at both sides of the lower portion of fan palm's main vein are equivalent to the outer ligulas of Gramineae. Because this structure of fan palm covers the nodes where a majority of the leaflets attach, it is incorrect to explain it with the simple structure of a simple outer ligula (which constitutes the upper end of the sheath's outer axis). This structure is very similar to the webbing structure on both sides of paspalum's branch spike axis (Fig. 5.4). That means the webbing structure on both sides of the branch spike axes of paspalum is also the node to which all of the branch spike axes of fan palm attach. However, they are also different. In Gramineae, this webbing is located above the branch spike axes, while in fan palm, it is formed below the branch spike axes. Nevertheless, this difference is not insurmountable in principle. Whether this secondary webbing structure is formed above or below a branch spike axis depends on the physiological conditions above or below the secondary growth; it is independent of a particular position above or below the axis. This kind of example is not uncommon in higher plants. For example, as we discussed before, the two axes on hooded-awn barley's glumes split back to back against each other.

Multiple leaflets mean multiple axes. The multiple vascular bundles formed by secondary differentiation all downwardly enter the petiole, causing the downward formation of multiple layers of evenly parallel vascular bundles in the petiole. These vascular bundles form a ring at the base of the petiole (i.e., the base of the leaf cleft) and enter the stem formed by the convergence of the lower portions of multiple axes. Parallelly with one another, these vascular bundle rings of numerous leaf clefts extend into the stem formed by multiple axes and connect to the vascular bundles below to form the whole vascular bundle network.

Although palm's primary radicle can be elongated and can regenerate branch roots, it has no laterally growable cambium, and thus it cannot be thickened and is limited in its absorption of nutrients and water. On the basal node of the stem, plenty of secondary roots grow, forming a fibrous root system unique to monocotyledons, which not only ensures the supply of nutrients and water to the large plant body but also guarantees physical support to the large body. These fibrous roots, like the branch roots formed on the radicle, are developed from secondary growing points outside the stele. The secondary growing points at the stem's base have a better nutrient supply than those on the radicle; they thus grow more vigorously and are thicker and stronger. In return, these thick and strong fibrous roots can also absorb more water, and the nutrients are dissolved in the water for consumption by the large plant body and have the ability to support the large plant body. Monocotyledons are different from dicotyledons. The size and strength of these organs in monocotyledons are shaped early during growth. They are unlike dicotyledonous trees, which have lifelong meristems, the cambia, and thus can be compensated by subsequent growth, making the roots and stems larger and thicker.

Wheat (*Triticum aestivum* L.) can be regarded as a representative of Gramineae. Figure 7.18 shows the structures of wheat in the early stage of germination. In the figure, h shows the whole picture when a wheat grain germinates to become a seedling. The young coleophyll emerges from the axilla of the second axial leaf cleft (the outer embryonic leaf). Three so-called radicles and basal coleorhiza are constituted by the structures in the lower portions of the first and second axes. The dedicated absorption organ, the scutellum, is the specialized first leaf cleft. The scutellum formed in the upper portion of the first axis differentiates relatively large vascular bundles to transport the nutrients that the scutellum absorbs from the endosperm. No differentiation of vascular bundles occurs in the second axis, the outer embryonic leaf. The coleophyll formed from the third axis in common wheat differentiates two vascular bundles, which aggregate together and interconnect with those of the axes above and below to form a seedling node, which constitutes the transportation system during the seedling stage.

From the transection of wheat's foliar organs shown in Fig. 7.19, we can see that the number of vascular bundles increases from one in a spine-like supernumerary

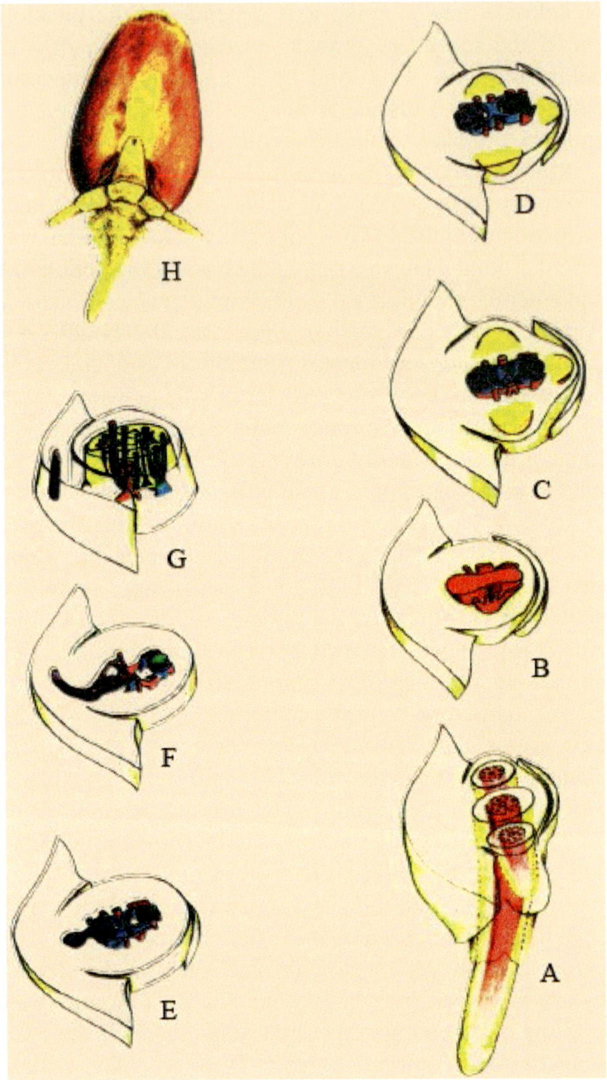

Fig. 7.18 The structure of wheat's young embryo. (**a**) The lower portion of the young embryo, showing the embryonic root sheath and the endosperm-absorbing embryonic leaf, which is composed of the first axis, and the scale and the three secondary radicles formed from the second axis; (**b**) the nodes formed by the convergence of the vascular bundles of the secondary radicle and those of the upper axis; (**c**, **d**) three secondary root primordia developed on the node; (**e**, **f**) the connecting structure of the secondary vascular bundles of the embryonic leaf with those of other axes; (**g**) the middle segment of the vascular bundle of an embryonic leaf, two secondary vascular bundles of the embryonic leaf sheath, and the relationship with the vascular bundles of the upper axis; (**h**) the appearance of a germinated wheat seed

spikelet to three and then to five in a flat supernumerary spikelet. The axes continue to enlarge and flatten and become hollow and then split open. Meanwhile, the number of vascular bundles increase to more than ten. The axes then differentiate into glumes, lemma, and palea. Because of the difference in volume and area proliferation, different and suitable numbers of vascular bundles are formed. The leaf sheath has the greatest number of vascular bundles, which is compatible with its large role in transportation. Vascular bundles are secondary. They are differentiated from primary parenchyma cells induced by the biosynthesis and flow path of

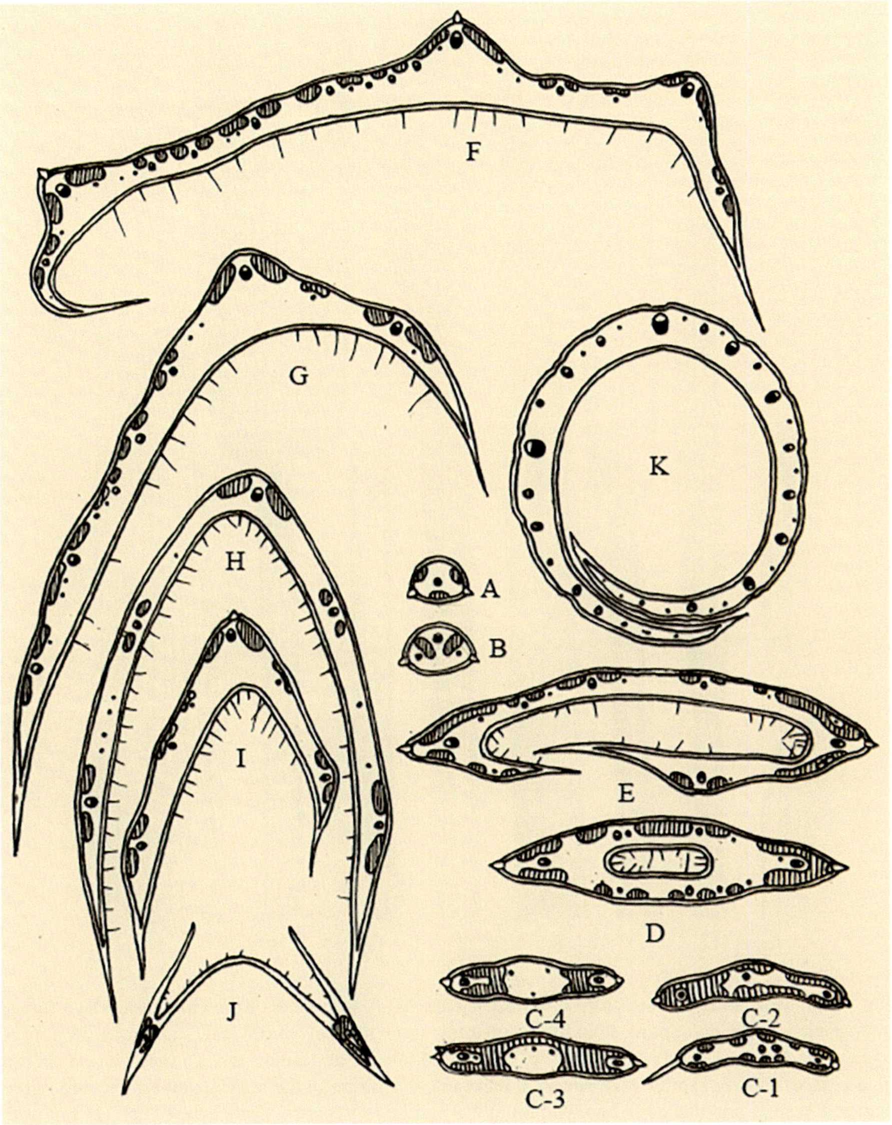

Fig. 7.19 Transections of wheat's foliage organs. (**a**) A spine-like supernumerary spikelet with a single vascular bundle, (**b**) an expanded spine-like supernumerary spikelet with three vascular bundles, (**c**) supernumerary spikelets flattened at four different degrees, (**d**) an expanded flattened supernumerary spikelet with a cavity, (**e**) a split flattened supernumerary spikelet, (**f**) a glume of a supernumerary spikelet with floras just being initiated, (**g**) a well-developed glume of a supernumerary spikelet, (**h**) a lemma, (**i**) an underdeveloped supernumerary spikelet lemma, (**j**) a palea, and (**k**) a leaf sheath

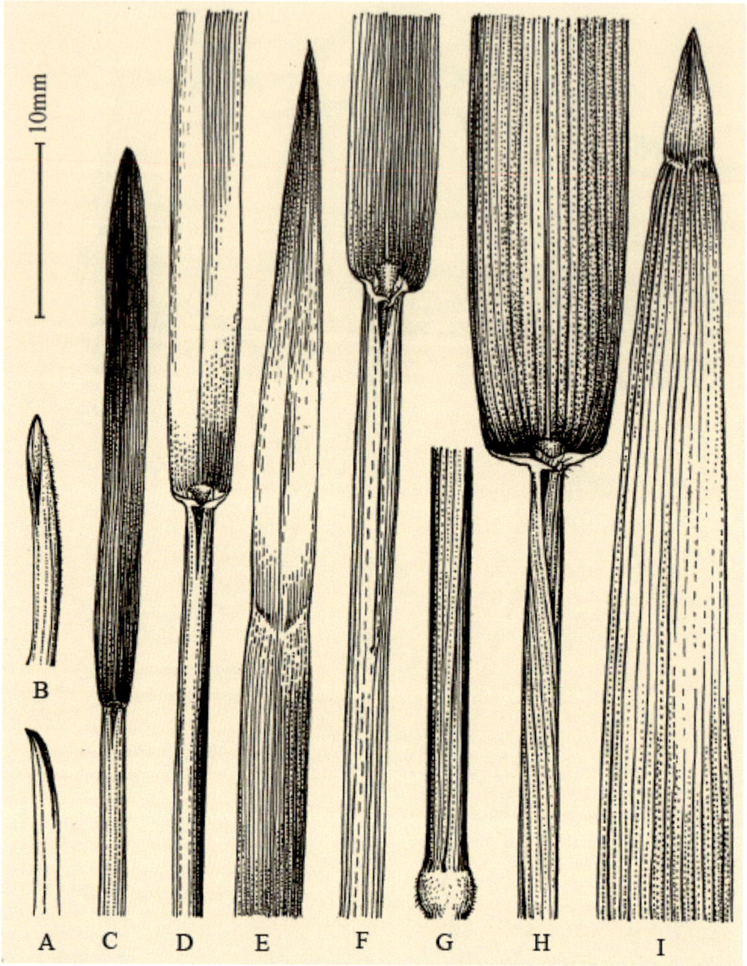

Fig. 7.20 Morphological structure of wheat leaves. (**a**) A side view of the coleoptile; (**b**) a three-quarter view of the coleoptile, showing an opening at the top; (**c**) a basal leaf, showing a very short opening at the top of the leaf sheath; no developed auricle and ligule and no constriction on the blade; (**d**, **e**) a lower cauline leaf, showing incomplete splitting of the leaf sheath; auricle and ligule start to develop, and there is a constriction at the middle to upper portion of the blade; (**f**) an upper cauline leaf, showing that the sheath does not completely split yet, and auricle and ligule are well developed; (**g–i**) a flag leaf, showing that the sheath has completely split, auricle and ligule are well developed, and a constriction is near the blade tip

gibberellic acid. They are physiologically inducive but not as primary as the telome theory believes.

Let us look at the structures of wheat leaves. Figure 7.20a, b shows half of the ventral and dorsal sides of the coleoptile, respectively. The coleoptile is composed of a hollow axis with a split upper portion and only two vascular bundles. For stem elongation and heading, wheat has to undergo vernalization. The basal leaf sheath splits only at the upper portion before stem elongation (Fig. 7.20c–e). The leaf structure of the initial basal leaf is also relatively simple; no constriction is visible on the leaf blade; auricle and ligula are not developed (Fig. 7.20c). The sheath of a relatively upper basal leaf is only split in the upper portion; the auricle and ligula begin to take shape; a constriction is located in the upper or middle part of the blade. A cauline leaf is still not a fully split sheath after stem elongation (Fig. 7.20f), full splitting only occurs on the flag leaf (Fig. 7.20g, h), auricle and ligula are fully developed, and the leaf tip has a constriction (Fig. 7.20i). The number of vascular bundles in wheat's sheath and blade increases as the leaf size increases. Observing such leaf structures must be done to compare them with those of hooded-awn barley. It is known from the evolution of the hooded awn of barley that auricles have been evolved from two protoglumes by dorsally merging into one, followed by splitting into two. An auricle is formed by merging two axes at the two nodes. The nodes of barley, wheat, and other Gramineae plants are actually formed by the evolution of two or more axes. The constriction is a flattening of several nodes. In other words, it is composed of two or more axes. This viewpoint is derived from observations on the evolution of the awn of hooded-awn barley. The awn is a compound node. There are three such compound nodes on wheat cauline leaves: one is the lowest collar (Fig. 7.21a–c); another one is the ring that connects the auricle, sheath, blade, and ligule (Fig. 7.22a, b); and the top one is the constriction. From the compound nodes on barley's hooded awn, one can see that these compound nodes are all built with three or more axes. According to the old school of morphology, the leaves of wheat are "single leaves." In fact, a wheat leaf is a compound leaf composed of more than ten axes, which is completely different from those true single leaves of lycopods, spike moss, pine, cedar, cypress, ephedra, and *Welwitschia*. For the stamen, however, the filament is an axis, and the anther is composed of two dorsally connected axes. The pollens (the reproductive generation gametophyte) are evolved from the pollen mother cells by meiosis within the anther axis (axilla) (Fig. 7.23). However, these meristematic parenchyma cell masses do not polarly differentiate into axes; each cell undergoes meiosis to form pollens, the life form of the next sexual generation. The formation of the female gametophyte is similar. In the uppermost portion of the floral organ, the enlarged hollow axis constitutes the ovary; the internally generated axis forms the seed coat. Within the seed coat axis, primordial cells on the newly generated axis develop into four gametophytic cells of the new generation by mitosis. One of the four then develops via three mitosis into a gametophyte composed of seven cells. One of the three upper cells develops into an egg and the other two into synergids; the one in the middle undergoes mitosis without cytoplasmic separation, developing into a large binucleic blastocyst; and the bottom three become podocytes. This is the constitution of the gametophytes of all seed plants, including wheat. The gametophyte parasitizes within the seed coat axis of the sporophyte (the previous generation) until the egg is fertilized and develops into a young embryo, the new generation of sporophyte. The blastocyst develops into endosperm after fertilization without independent living.

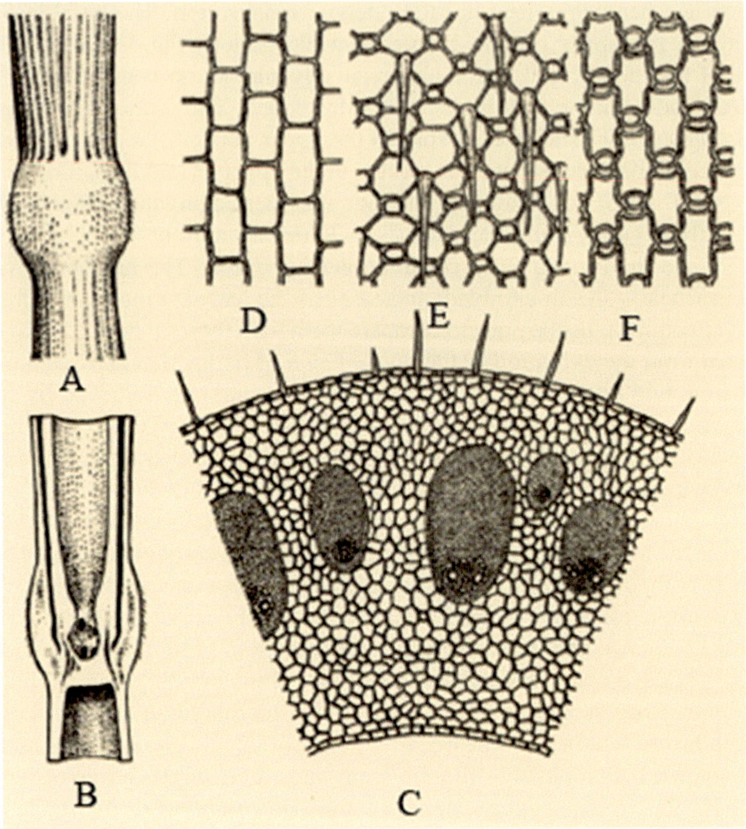

Fig. 7.21 The structure of a common wheat node. (**a**) The upper portion of the internode, the node, and the lower portion of a flag leaf's sheath, showing the collar; (**b**) a longitudinal section of the upper end of the internode, the node, and the lower portion of a flag leaf's sheath, showing the internode and the fattened thick collar; (**c**) a transection of the collar; (**d**) the internode's epidermal cells; (**e**) the collar's epidermal cells; (**f**) the sheath's epidermal cells

The structure of rice (*Oryza sativa* L.) is basically consistent with that of wheat and barley. Its first axis forms the endosperm-absorbing scutellum and radicle (Fig. 7.24c). The axillary second axis forms the hollow coleoptile with an opening at the upper end. For the cauline leaves, the sheath above the collar splits open all the way to the collar; auricles and ligules are well developed, except that the outer ligules have degenerated and thus are not visible (Fig. 7.24a), but the constriction is visible (Fig. 7.24b, indicated by an arrowhead). After learning how the same structures of wheat and barley have evolved, these structures of rice should also have evolved from axes in the same way.

Among the monocotyledons, Araceae and Gramineae are quite different in their organ morphologies and constitutions. Except for the spathe under the florescence, which is a typical single-split axis, its green leaves are compound leaves with

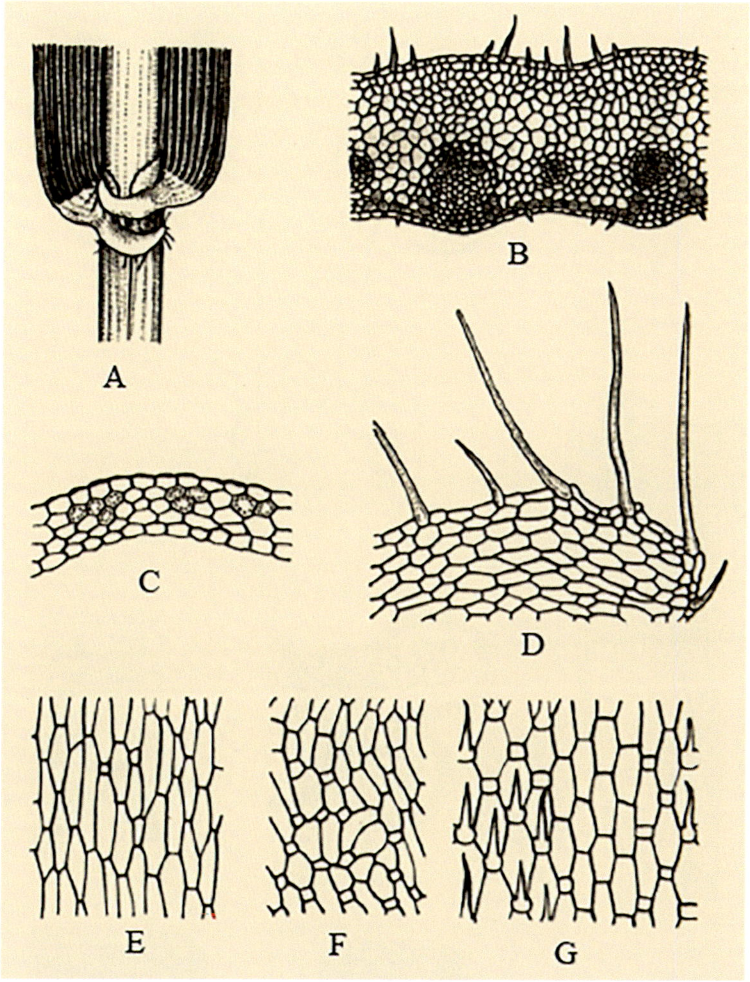

Fig. 7.22 The structure of wheat's auricle and ligule. (**a**) An amplexicaul leaf sheath and the lower portion of the blade, showing morphological structures of the ligule, auricle, and ring; (**b**) the transection of a ring; (**c**) the transection of a ligule; (**d**) a part of a ligule, showing long hairs along the edge; (**e**) the epidermal cells of the ligule; (**f**) the epidermal cells in the transition zone between the ring and the ligule; (**g**) the upper epidermal cells inside the ring

reticulate veins formed from multiple axes and thus are different from those of most monocotyledons. Figure 7.25 shows the compound leaves of leopard palm (*Amorphophallus rivieri* Durieu). Layer after layer of multiple vascular bundle rings are visible from the transections of different parts of the leaf; the aggregation of multiple axes is shown; reticulate venation is similar to that of dicotyledons. Such structures are not explainable by the telome theory because the telome theory believes that veins are branches and branches are reticulated and composed of

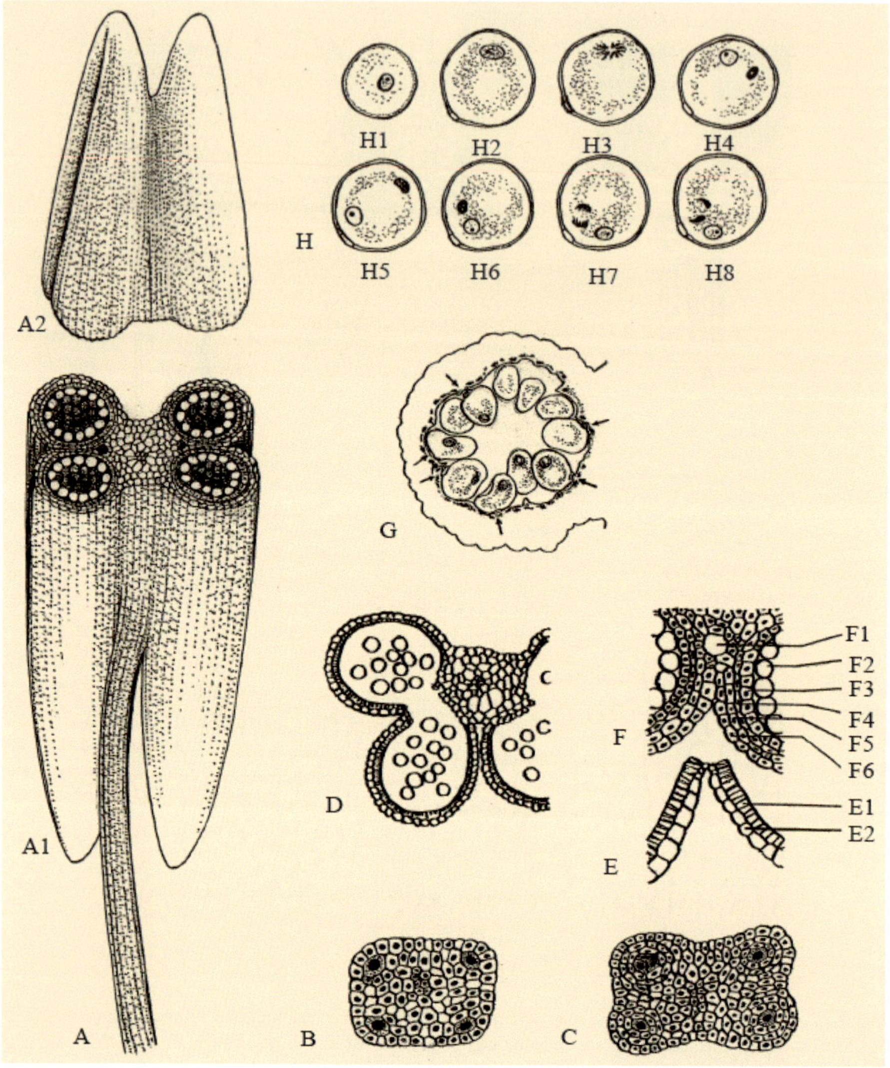

Fig. 7.23 The structure of wheat stamen. (**a**) The structure of a near-mature stamen: the transection of the anther, showing the inner structure and external morphology of the middle and lower portion (a1) and the external morphology of the upper portion of the anther (a2); (**b**) the transection of a young anther, showing the archesporial cells with relatively large nuclei and dense cytoplasm formed from the subepidermal cells of the four edges; (**c**) the transection of a young anther that develops further, showing the endothelial cells formed from archesporial cells by periclinal division in the four edges and the pollen mother cells formed at the center; (**d**) the transection of a mature anther, showing the pollen grains inside; (**e**) the mature anther wall: the fibrous layer (e1) and epidermis (e2); (**f**) the transection of a part of a near-mature anther, showing the structure of the anther wall, stomium (f1), pollen grain (f2), tapetum (f3), middle layer (f4), fibrous layer (f5), and epidermis (f6); (**g**) the transection of a medium-term anther, showing the arrangement of pollen grains in the anther; the arrowhead points to the position of the aperture; (**h**) development of a pollen grain: the uninuclear stage (h1 and h2), the binuclear stage (h3 and h4), the second nuclear division in the male gamete (the division of the sperm nuclei) (h7), the trinuclear pollen grain (h8). ((**d**–**f**) are adapted from Percival 1921; (**g**, **h**) are adapted from Kihara 1937)

Fig. 7.24 *Oryza sativa* L. (**a**) Collar, sheath, auricle, ligule, and the lower portion of the blade; (**b**) the blade, showing the constriction (pointed by the arrowhead); (**c**) the longitudinal section of a germinated rice grain

multiple layers, which cannot be justified. In the first year, leopard palm develops one compound nutrient green leaf, which stays independently above the ground, up to 40–80 cm. The blade is trifarious, reticular, and palm like and is 30–70 cm wide. The base of the petiole still clearly shows a split axis structure. The multiaxis, multisystem, down-extending vascular bundles are arranged in multiple layers (rings), yet the original shape of the petiole split axis remains unchanged. The multiple years of intercalary growth of the lower axial portions have made the stem formed by the aggregation of the lower axial portions become an oblate tuber, from which multiple secondary fibrous roots grow. This is the organ morphological constitution that is unique to Araceae.

Yam [*Dioscorea esculenta* (Lour.) Burk] is a monocotyledon with single-split axes; thus, its phyllotaxy should be alternate, and it actually is alternate. However,

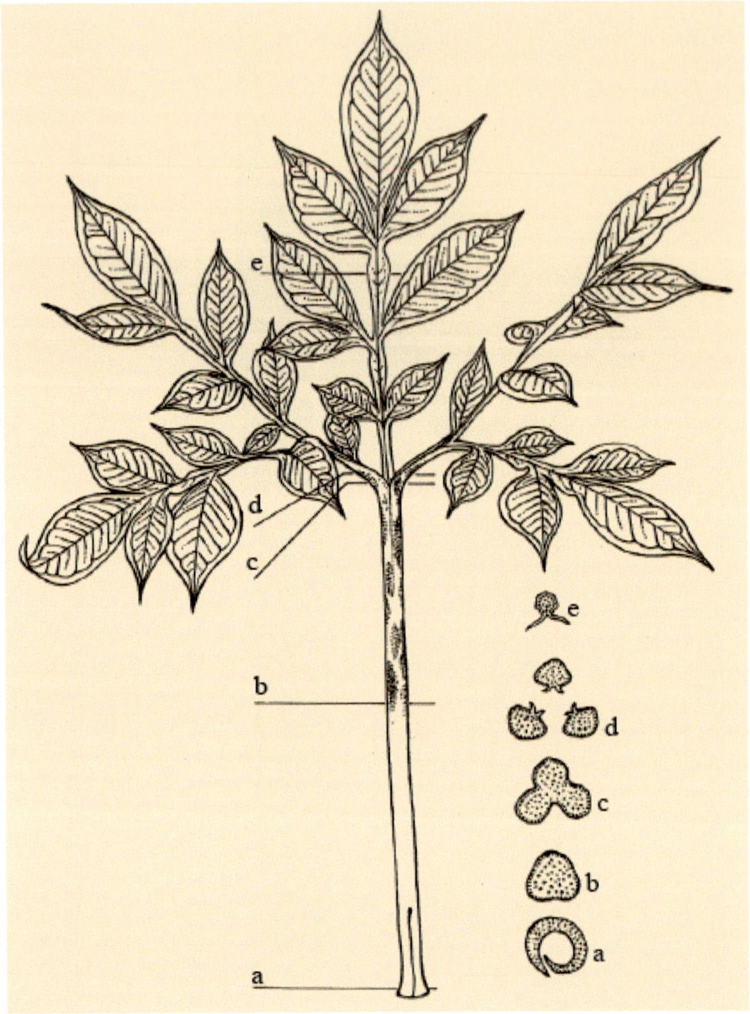

Fig. 7.25 A leaf of *Amorphophallus rivieri* Durieu and a transections of its petiole

opposite phyllotaxy often appears above the middle nodes. Obviously, the opposite phyllotaxy is formed by alternative aggregation. This seems to be in line with the concept of the telome theory that alternative telomes are aggregated to form opposite phyllotaxy. However, as discussed before, there has been sufficient evidence to prove that leaves are not webbed branches but split axes. Therefore, the petiole and stem structures of leopard palms also directly indicate that leaves are split axes, not webbed branches (Figs. 7.26 and 7.27). The evolution of opposite phyllotaxy from alternate phyllotaxy is due to the almost simultaneous development of the upper and lower split axes without the formation of internodes. The phyllotaxy has the morphology of aggregation, aggregating the alternate to form the opposite.

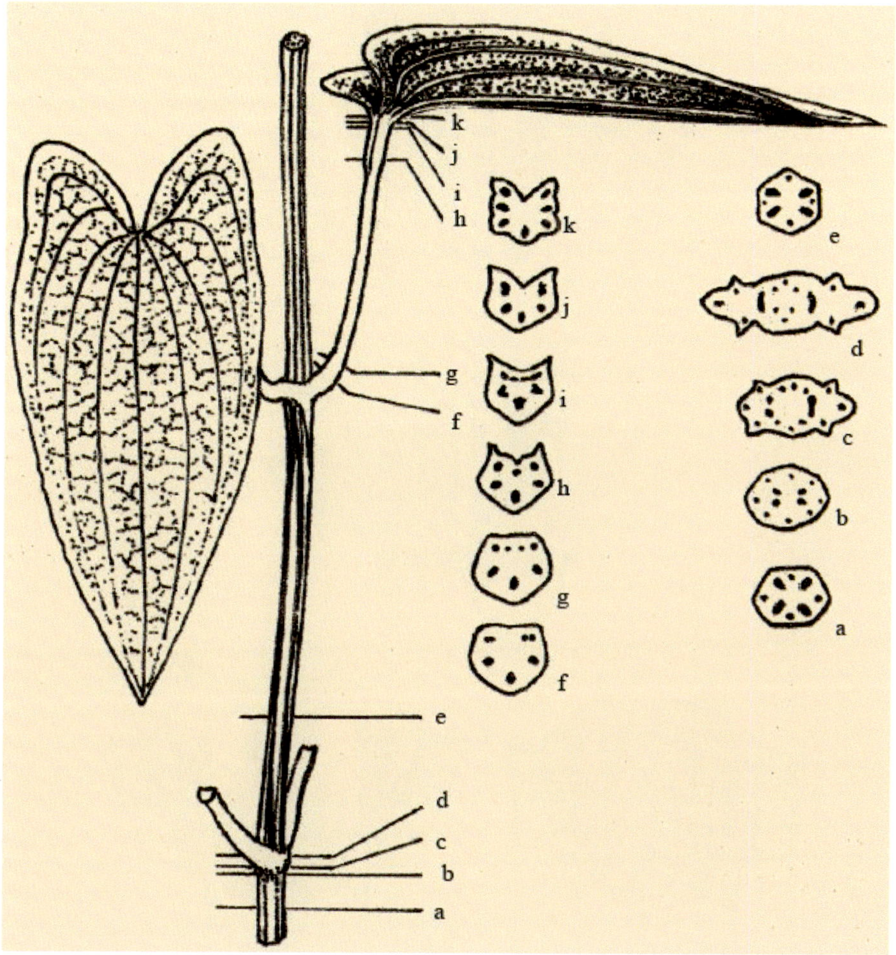

Fig. 7.26 The opposite leaves on the stem of *Dioscorea esculenta* (Lour.) Burk and their transections. (The alphabets in the figure are for illustration of the relative position only)

However, it is not an aggregation of the telomes but an aggregation of individual alternative leaf clefts.

Dioscorea has plants with foliar structures with one to three leaflets, such as the leaves of *Dioscorea biformifolia* Péi et C.T. Ting. This fully proves that yam leaves are heart-shaped compound leaves formed by webbing multiple leaflets together and, thus, are constructed from multiple axes. From the transection shown in Fig. 7.26g–i, you can see the vascular bundle from the upper axis on the ventral surface of the leaf cleft. The webbing that we observed before on the spike axes of crabgrass and paspalum of Gramineae is reflected on yam's leaves. However, it is not the so-called branch webbing under the telome theory but a heart-shaped compound leaf formed by webbed leaflets. Vascular bundles are secondary structures, which is a proven fact

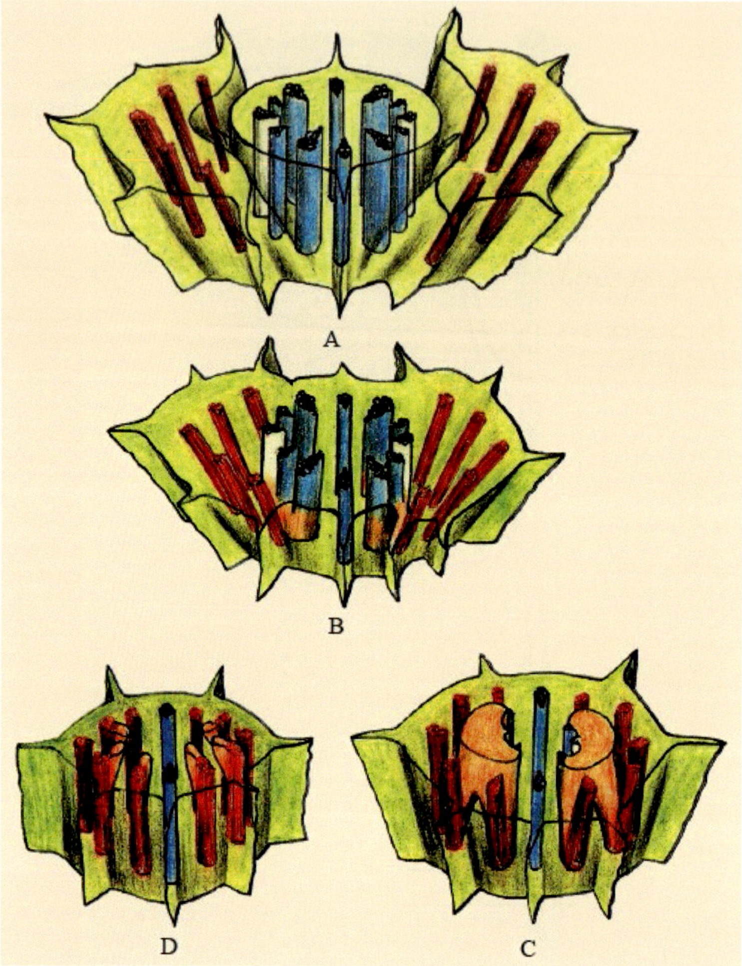

Fig. 7.27 The vascular bundles in the node of the opposite leaves of *Dioscorea esculenta* (Lour.) Burk. (Colors indicate different axial relationships. (**a–d**) are corresponding to a–d in Fig. 7.26)

in experimental biology. Even without mentioning the conclusions of experimental biology, the reticulate venation of yam is hard to be explained by the telome theory. The examples of yam have shown that there are approaches of webbing and aggregation in the organogenesis of higher plants. However, they are not the so-called telome webbing and aggregation under the telome theory but those between leaf clefts. Nodes are the place where vascular bundles of the leaf clefts formed by different axes join together. Taking Fig. 7.27 as an example, white color shows the downward-extending vascular bundles aggregated at the uppermost axis; blue color shows the vascular bundles of the upper axis; orange shows the vascular bundles that have been aggregated to form the upper node and are extending downwardly, like the white vascular bundles of the uppermost node; and red shows the circumstance that the vascular bundles of two aggregated leaf clefts with five vascular bundles each enter the stem segment through the petiole. These vascular bundles are all secondary structures induced by auxin-like active substances.

7.2.2 *Dicotyledons*

Green bean (*Phaseolus vulgaris*) is a dicotyledon. Its first pair of leaves are opposite, and all leaves above are alternative but the stipules on each node are all opposite. From the perspective of its compound nodes, the first, second, and third nodes are displaced from each other by 90° from top to bottom (Fig. 7.28a–c); the alternative compound leaves above the third node are displaced from each other by 180°. The stipule is the first axis of the compound leaf; its lower portion combines with the lower portions of other axes to form the stem; its upper portion splits into stipules (Fig. 7.28a, b). The green leaves above the stipules are formed by the axes regenerated multiple times on the stipules. Cotyledons are simple leaf clefts, the lower portion of which form the radicle. Before developing into a mature seed, the nutrients in the endosperm are absorbed and transported by the cotyledons and stored in them so they become endosperm-less seeds. After germination and emerging from the soil, the cotyledons provide the nutrients needed for seedling growth, and vascular bundles in the cotyledons develop into a web (Fig. 7.28c). These reticulated vascular bundles cannot be explained by the telome theory. The reticulated vascular bundles formed in each cotyledon converge into four bundles, and the eight bundles of the two cotyledons further converge into four bundles after they enter the radicle and finally aggregate to form the quadrangular stele of the radicle (Fig 7.28c–i).

Pea (*Pisum sativum* L., Fig. 7.29) and broad bean (*Vicia faba* L., Figs. 7.30 and 7.31) are similar in organ construction. However, the cotyledons of pea, as a nutrient storage organ, do not emerge from the soil. Like the green bean, the vascular bundles in the cotyledons of pea are reticulated. All leaves above are alternately displaced from each other by 180°. That is to say, the two leaf clefts are opposite. One leaf cleft is mature and thus fixed first, and the other is still in the stage of growth, development, and elongation. As a result, the immature leaf cleft rises above the mature one to an alternative position. Therefore, the phyllotaxy becomes opposite-alternate. The compound leaf on the second axis becomes scaly and dichotomic, and the new axis grown between forms a small cusp (Fig. 7.29b, c). The leaf clefts on the nodes of the third axis and the above develop into two stipules. New axes grown on the stipules form compound green leaves. Their leaf veins are all reticulated. All these are negations of the telome theory. Pea and broad bean are different from green bean in that their second axes do not form opposite green leaves. Instead, the two leaf clefts are displaced from each other to form two oppositely positioned scaly leaflets. Each of the leaf clefts is 180° offset to form pinnate alternate compound leaves. As a result, the entire plant has a distichous opposite-alternate phyllotaxy.

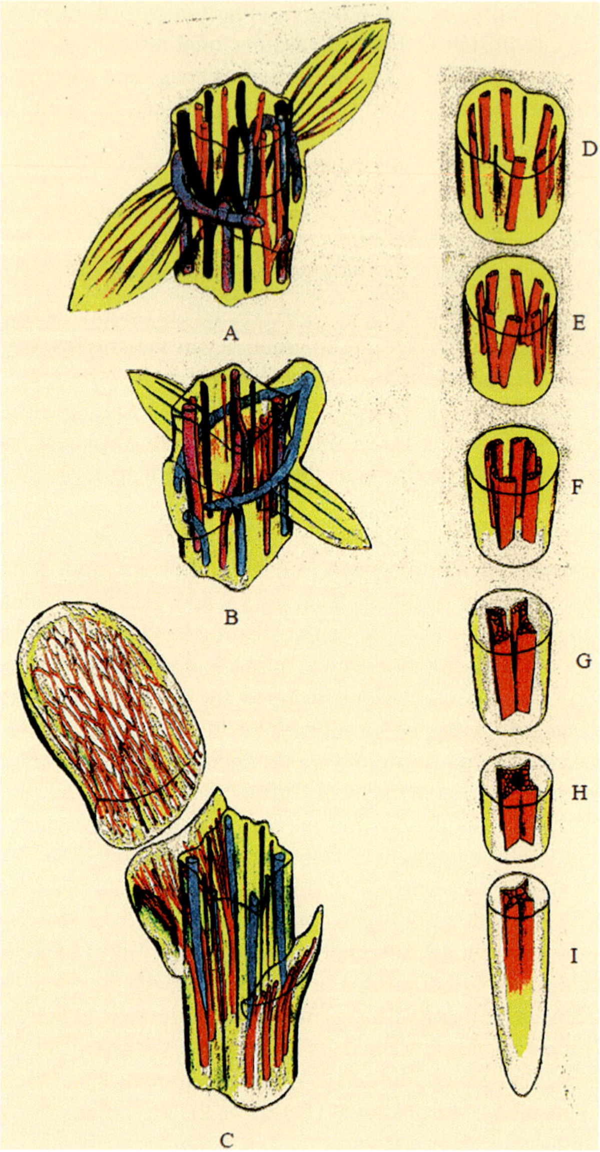

Fig. 7.28 The vascular bundle system of green bean seedling, showing the interconnection of the axes (the colors are not original; they are here to distinguish different axial systems). (**a**) The third axis, an alternate leaf node, showing the relationship of the vascular bundles; (**b**) the second axis, an opposite leaf node, showing the relationship of the vascular bundles; (**c**) the first axis, the cotyledon node, showing the relationship of the vascular bundles; the axis above has only four vascular bundles (blue) and are connected with those of the first axis; (**d**) the hypocotyl, convergence of the web-like vascular bundles of the two cotyledons into four bundles when entering the hypocotyl; (**e**–**i**) the eight vascular bundles aggregated in the two cotyledons are further combined into four bundles, which form the radicle vascular bundles

The leaflet on the tip of broad bean's pinnate compound leaf is often trumpet shaped (Fig. 7.32). This trumpet-shaped blade structure indicates that it is a semi-split axis. Such trumpet-shaped, semi-split axis is also often found in other higher plants, such as the aforementioned sheath leaves of wheat, barley, and rice; the awn of hooded-awn barley; and the semi-split supernumerary spikelets of wheat. They all have such structure. This structure is also found on the compound leaves of

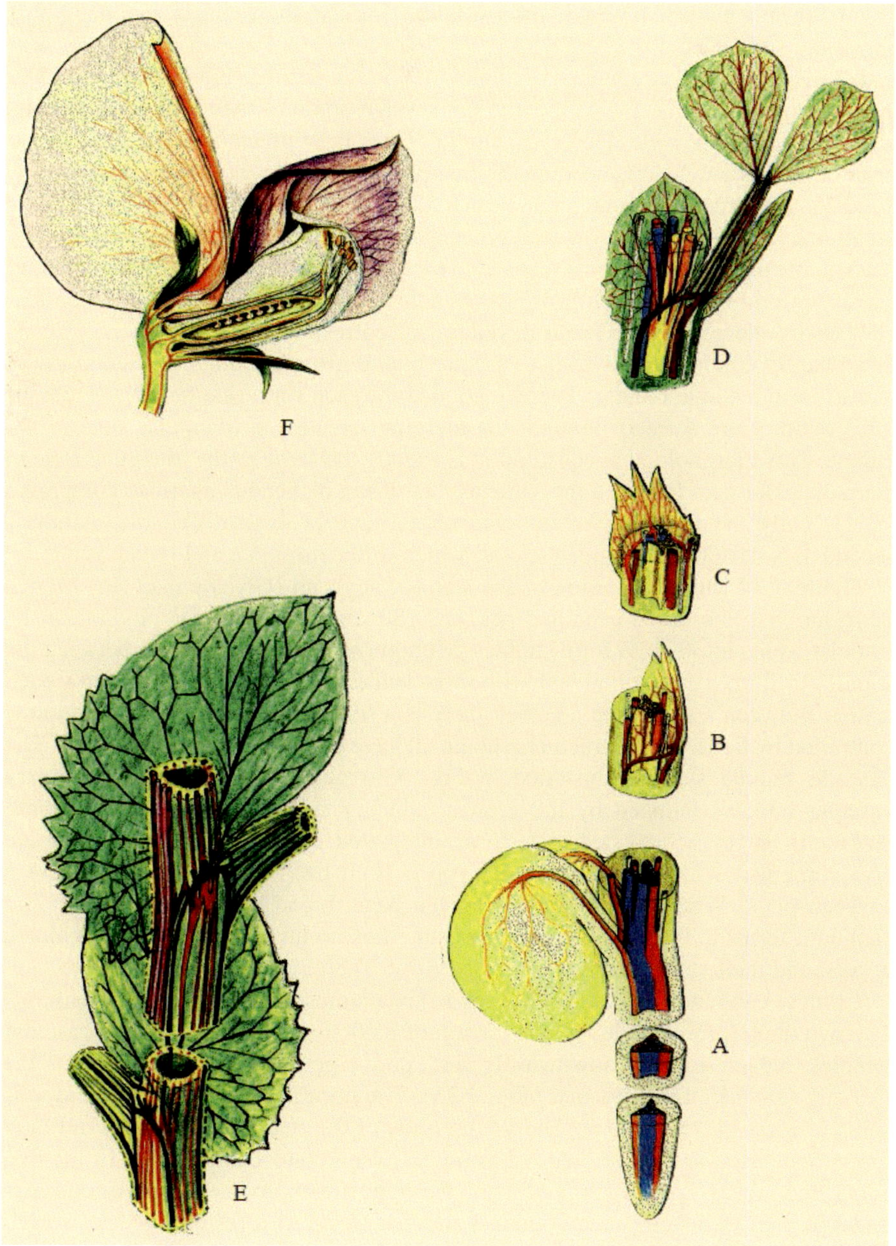

Fig. 7.29 The vascular bundle system of pea (*Pisum sativum* L.), showing the interconnection between axes (the colors are not original; they are here to distinguish different axial systems). (**a**) The first axis forms the cotyledons, hypocotyl, and radicle, and the secondary vascular bundle webs converge into two bundles when entering the hypocotyl and are, together with those forming the upper axes, aggregated into one system in the radicle; (**b**) the interconnection of the vascular bundles in the second axis; (**c**) the interconnection of the vascular bundles in the third axis; (**d**) the interconnection of the vascular bundles in the fourth axis; the vascular bundles developed in the leaf blades are meshed into a web; (**e**) the interconnection of the vascular bundles in the upper axis; the

cruciferous plants, which we will discuss later. This also shows the commonality that leaves are split axes.

Like legumes, the cotyledons of alfalfa (*Medicago sativa*) (Fig. 7.33), when they emerge from the soil, serve as the nutrient storage organ of the seedling to meet its nutrient needs for growth. However, alfalfa differs from green bean, pea, and broad bean in that its cotyledons become the first pair of green leaves for photosynthesis. Alfalfa is the same as pea and broad bean in that one of the leaf clefts on the first axis preferentially develops to form an alternate phyllotaxy. It differs from pea and broad bean in that the leaf clefts on its second axis are not scaly but develop into compound leaves with three leaflets. The vascular bundle in this preferentially developed leaf cleft also undergoes preferential development correspondingly, so that it extends downwardly and is merged with a vascular bundle from the other cotyledon of the pair to form a single bundle. The two cotyledons each form one bundle. A total of three bundles are formed through this merging, creating a triangular stele in the radicle. This morphological constitution is exactly the same as the triangular stele in the radicles of pea. It is also the same as that of the distichous alternate compound leaves. Only the functions of the cotyledons have evolved differently. Alfalfa's second axis develops into green leaves, which differ from pea and broad bean.

Figure 7.34 shows a portion of the stem of soybean (*Glycine max* (L.) Merr.). Soybean, just like broad bean and pea, has 180° opposite-alternate phyllotaxy. Its leaf cleft splits into two to form stipules. Multiple secondary axes grow between the stipules to form a compound leaf with three leaflets with a scaly stipel below each leaflet. Between a stipel and a leaflet, there is a succulent petiole, whose motion is controlled by light. These structures should all have evolved from multiple axes. The vascular bundle system illustrated in Fig. 7.34 also shows that the secondary vascular bundles induced by the growth-inducing substances generated in these leaf clefts are interconnected to form an integrated conduction network; they are also combined with other organs developed from each axis into one to form a soybean plant. Whether it is soybean, green bean, broad bean, pea, or alfalfa, no cambium forms in their vascular bundles, and thus no increase of xylem or phloem occurs, and their stem cannot continually grow. Therefore, they become herbs.

Chinese cabbage (*Brassica chinensis* L.) is a representative plant of Crusiferae. The two dichotomic cotyledons on the top of its first axis each forms five vascular bundles, which extend downwardly and converge together to constitute the two-edge stele in the hypocotyl (Fig. 7.35). The upper portion of the second axis splits to form two euphyllae. The young buds of the two euphyllae each initially forms one vascular bundle, which has not yet been connected to those of the first axis (Fig. 7.35d). As growth continues, the opposite euphyllae of the second axis undergo uneven development, one fast and one slow. As a result, one leaf is aged and the other rises above to become alternate. On the third axis and the above, this

---

Fig. 7.29 (continued) vascular bundles developed in the leaf blades are meshed into a web; (**f**) the longitudinal section of a flower, showing the interconnection of vascular bundles in the floral organ

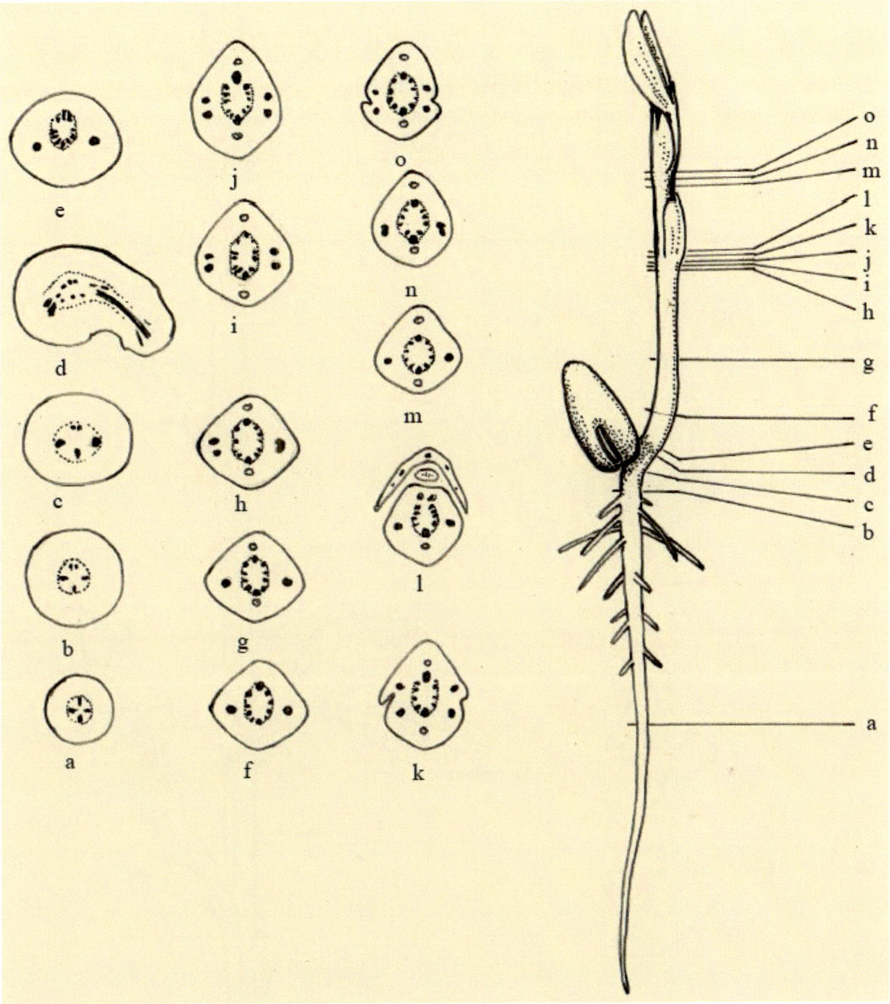

Fig. 7.30 The vascular bundle system of broad bean, showing the interconnection of axes; a: the transection of radicle, b: the transection of the transition portion between the hypocotyl and radicle, c: the transection of radicle, d: the transection of the cotyledon axis, e–g: the transections of the first internode, h, i: the transections of the second axis, m–o: the transection of the third axis

uneven development continues, causing a distorted rotation of the alternate phyllotaxy. Meanwhile, axes grow on the euphylla clefts to form composite simple leaves (Fig. 7.36b–e). The so-called composite simple leaves develop in a way that axes grown on an existing leaf cleft are flattened to form single leaves and these leaves are then webbed together to become a simple leaf. Among its leaf organs, cotyledons, calyxes, and petals are developed from a single axial leaf cleft (Fig. 7.36a, f, g).

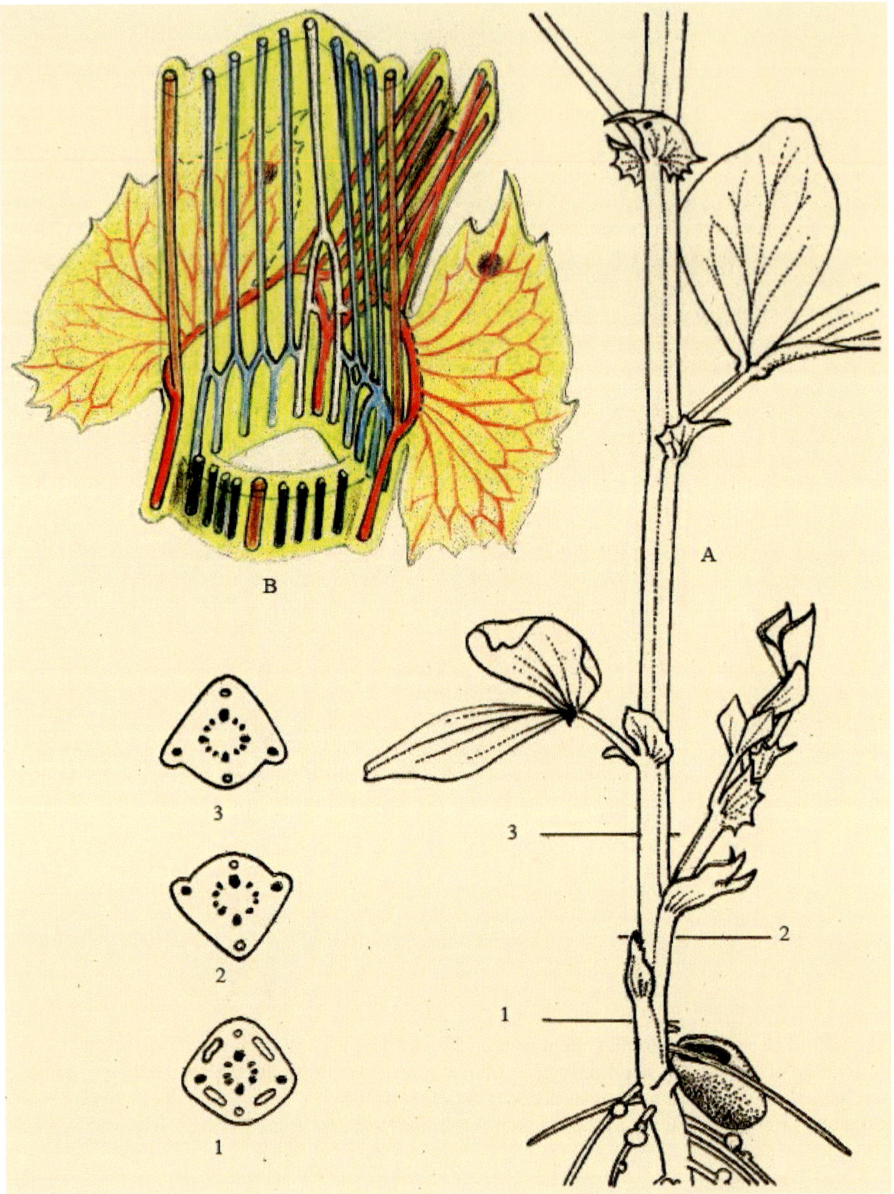

Fig. 7.31 The vascular bundle system of broad bean, showing the interconnection of the axes (the colors are not original; they are here to distinguish different axial systems). (**a**) The whole plant: (1) the transection of the first axis, (2) the transection of the second axis, (3) the transection of the third axis. (**b**) The interconnection of the vascular bundles of the upper and lower axes within the node

Fig. 7.32 A compound leaf of broad bean, showing the trumpet-like structure of the top leaflet (arrow-headed)

The multiaxial structures of the composite simple leaves of Chinese cabbage constitute a giant complex composite simple leaf system, resulting in an interconnected web of the complex multiple veins (Figs. 7.36 and 7.37). Although the cotyledons, calyxes, and petals are formed from single axes, their vascular bundle systems are still interconnected like a web (Fig. 7.36a, f, g).

A composite simple leaf can have more than ten vascular bundles; they extend from its petiole into the stem and are interconnected there (Fig. 7.36). Figure 7.37 shows how the veins are interconnected within a cauline leaf. The stem is a secondary structure formed by the aggregation of the lower portions of multiple axes. The vascular bundles differentiated in the leaves extend down into the stem where they interconnect with each other to make the entire plant's conduction tissue into a unified system and thus to make the secondary structure, the stem, a unified organ. The formation and development of the secondary xylem in the vascular bundles that feather strengthen the stem to be a solid whole. However, do not forget that stems do not exist in the early stage of growth and development. It is the lower portions of the axes united by the secondary vascular bundles. It is a union in itself. It is the interconnection of the secondary vascular bundles extending downwardly from each leaf clefts that makes them unite together and the subsequent xylem development that makes the stem a unified organ. The illusion of this union has deceived many botanical researchers because they never regard the stem as a subject in terms of its occurrence and development. This is the crux of why people mistakenly believe that the stem is primary. The stem is not an axial structure; it is a union of the lower portions of multiple axes. It is equal not to the sporophyte of hornworm but to a union of the lower portions of multiple hornworm sporophytes. This is a key point that the telome theory scholars fail to recognize.

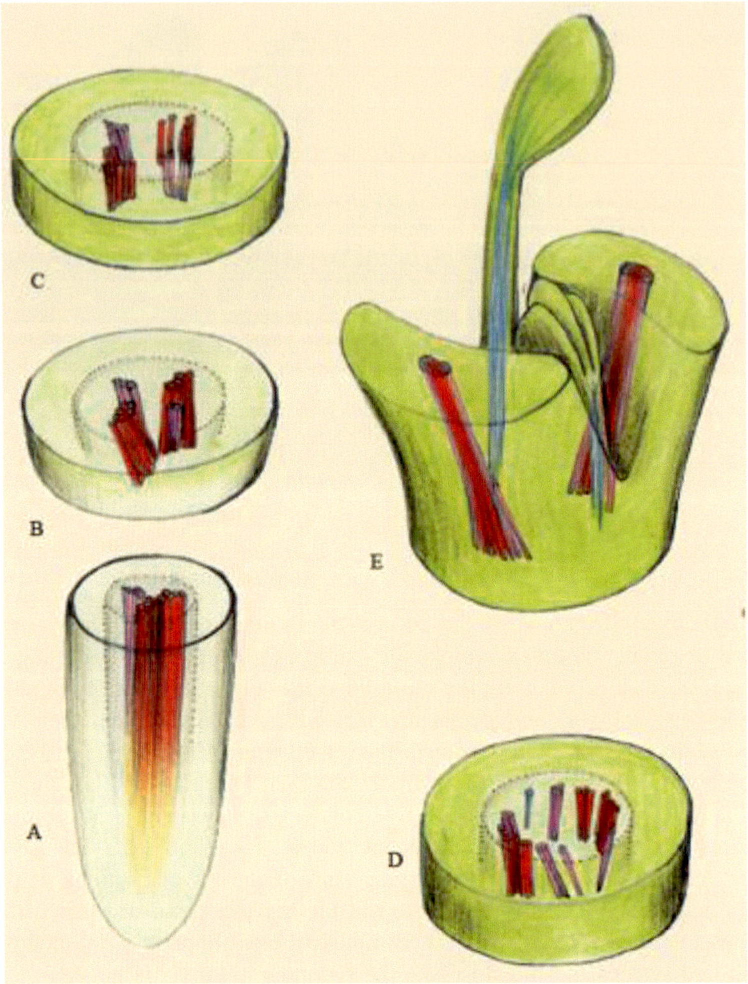

Fig. 7.33 The vascular bundle system of alfalfa (*Medicago sativa*), showing the interconnection of the axes (the colors are not original; they are here to distinguish different axial systems). (**a**) Radicle, (**b**) the transitional zone between radicle and hypocotyl, (**c**, **d**) hypocotyl, (**e**) cotyledon node

It is very natural that auxin-like products flow to form a network in a simple leaf formed from multiple secondary axes by webbing. From the perspective of the multiple secondary axis theory, therefore, the formation of the reticular venation of Chinese cabbage (Fig. 7.37) by secondary differentiation is an inevitable result. The telome theory cannot explain this because it regards a vein as a branch.

Mustard (*Brassica junceae* (L.) Czern et Cross.) is a relative of Chinese cabbage. They belong to the same genus. Its composite simple leaf is basically identical to that of Chinese cabbage. The minor difference is that some of its varieties have pinnate composite simple leaves (Fig. 7.38). What is introduced here is that the regenerated

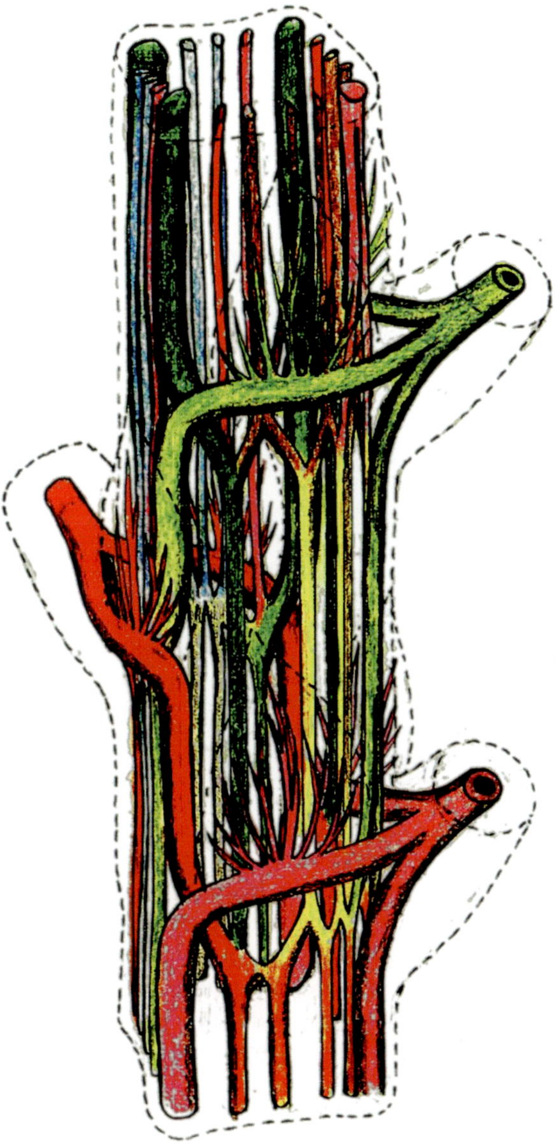

Fig. 7.34 The vascular bundle system of soybean, showing the interconnection of the axes. (The colors are not original; they are here to distinguish different axial systems)

axes in this specimen show the whole process that the secondary axes split and are flattened, which is similar to those observed on wheat's supernumerary spikelets (Fig. 7.38(1)–(3)). This kind of structure appears on wheat's supernumerary spikelets and the leaves of broad bean, and now it appears on mustard's leaves. The appearance of such unsplit new axes and the semi-split and fully split leaf cleftlets on mustard's composite simple leaves prove that mustard's compound simple leaves are composed of multiple axes. Webbing multiple axes together makes a pinnate

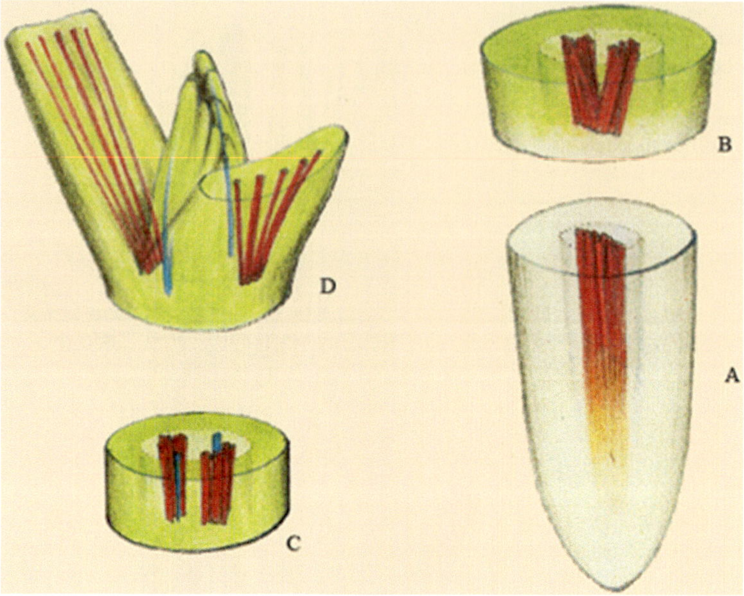

Fig. 7.35 The vascular bundle system of a Chinese cabbage (*Brassica chinensis*) seedling (the colors are not original; they are here to distinguish different axial systems). (**a**) Radicle; (**b**) the transition zone between the radicle and hypocotyl; (**c**) hypocotyl; (**d**) cotyledon node; the vascular bundles of the first axis (red) start differentiation at their bases, developing five bundles upward into the cotyledons while downward into the radicle to merge into one there. The vascular bundles in the euphyllae (blue) have not yet connected with those of the first axis

composite simple mustard leaf become a composite simple leaf, such as a Chinese cabbage leaf. This demonstrates the whole process of how to construct a composite simple leaf from multiple axes. Why does webbing occur? In the early stage of their differentiation, these axes grow together and develop together. As a result, their neighboring cells are connected together to form a union. After flattening, they become such a special form of construction, the "webbing."

Figure 7.39 shows the pinnate compound leaf of tomato (*Lycopersicon esculentum* Mill.), which is structurally similar to mustard's pinnate composite simple leaf. The difference is that tomato's pinnate leaflet develops a petiolule, and the petiolule is connected with the petiole. This kind of connection is also a kind of node construction. Buds can also form in the leaflet axillae of the node (Figs. 7.40 and 7.41). If we do an experiment to remove all other bugs from a tomato plant, axillary bugs will emerge from the leaflet axillae on these compound leaves (Fig. 7.40) and grow into branches. However, the petiole of the tomato leaf that develops the new branch always has a split axis structure. This will never be changed. This is similar to the aforementioned leaf of Japanese climbing fern, which can grow one section after another up to 4–5 m long, yet its petiole is still an axial structure. This is because this petiole (split axis) is no longer able to be aggregated with other axes to form a stem, and thus it can only be a split axis. This, in

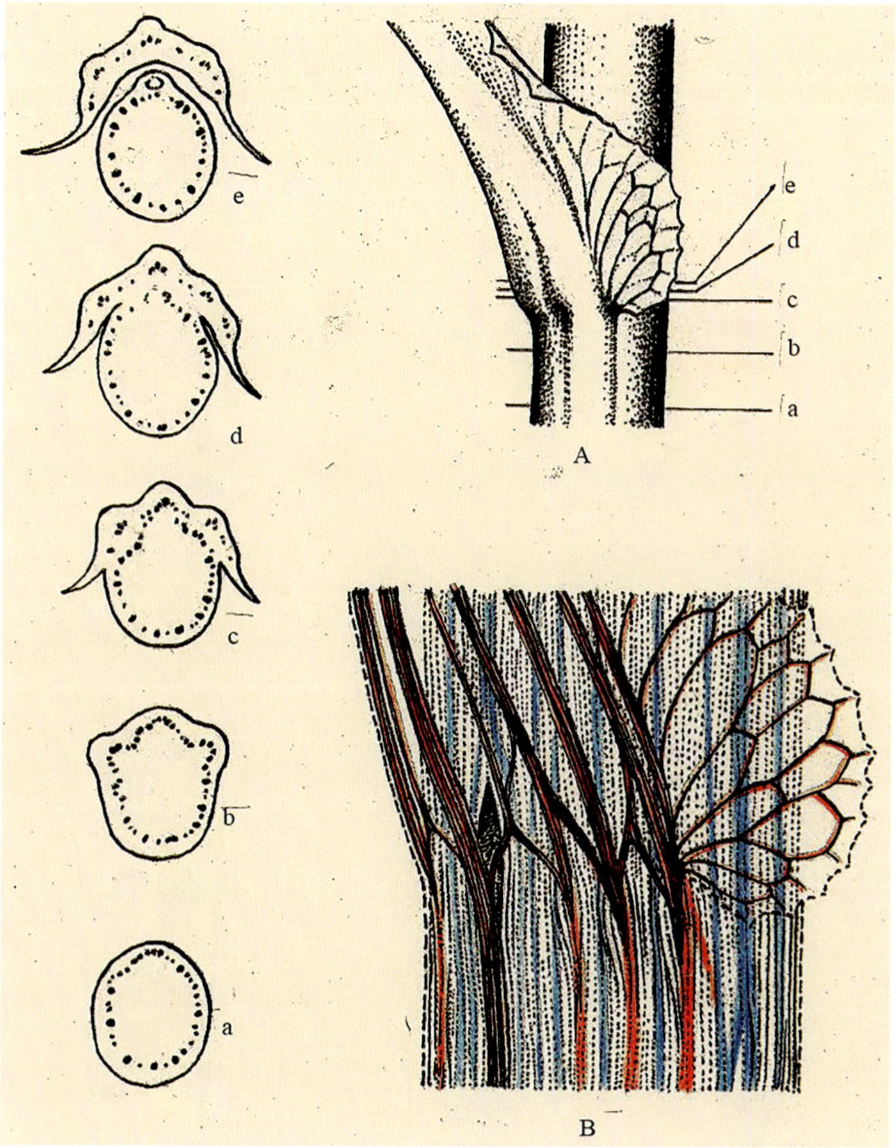

Fig. 7.36 The vascular bundles of Chinese cabbage (the colors are not original; they are here to distinguish different axial systems). (**a**) A segment of a stem, showing the positions of each transection; a–e are continuous transections, showing the process that vein vascular bundles enter into the stem to unite with the vascular bundles of the upper axis; (**b**) a perspective view of vascular bundles entering the stem and uniting with those of the upper axis

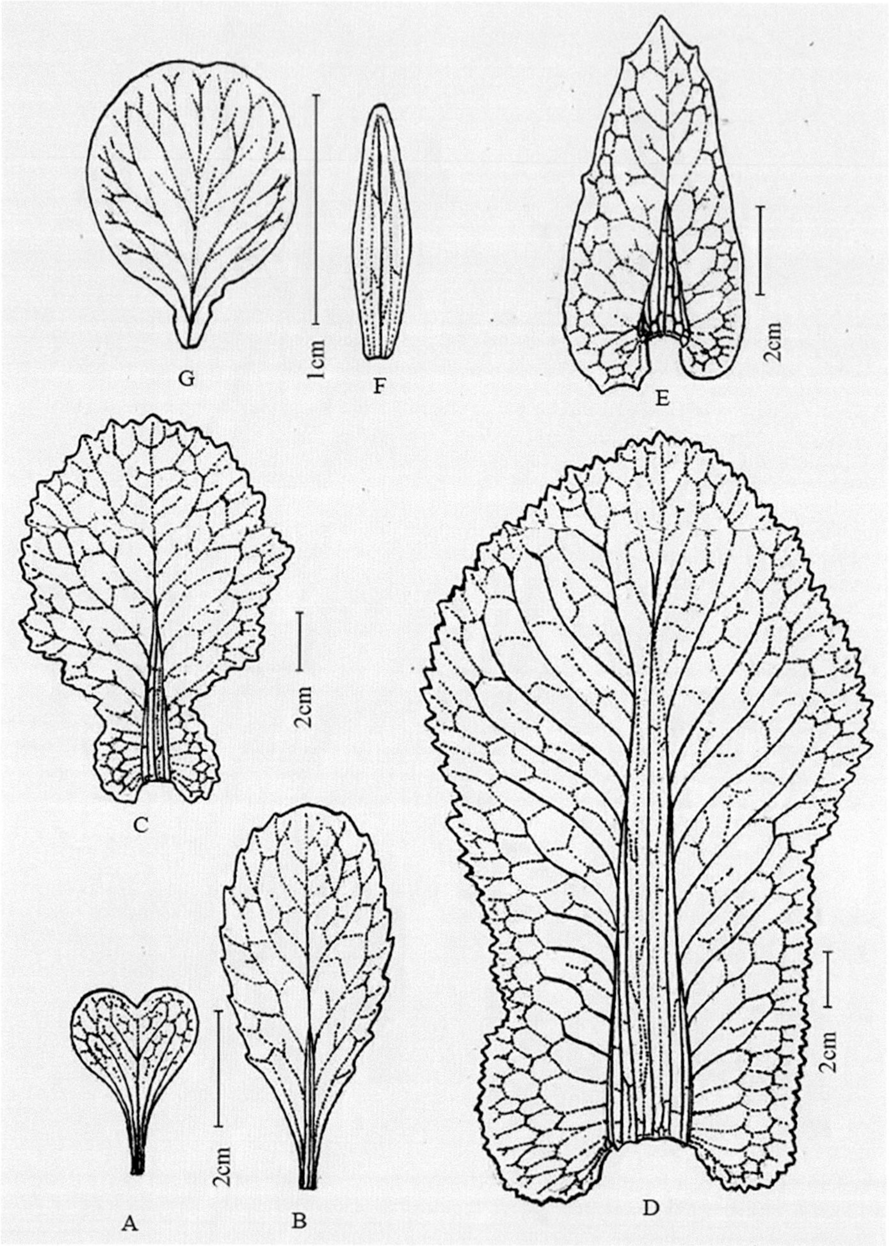

Fig. 7.37 The vascular bundle system of the phyllomes of Chinese cabbage. (**a**) A cotyledon, (**b**) the first euphylla, (**c**) a basal leaf, (**d**) a basal curve leaf, (**e**) a cauline leaf, (**f**) a calyx, (**g**) a petal

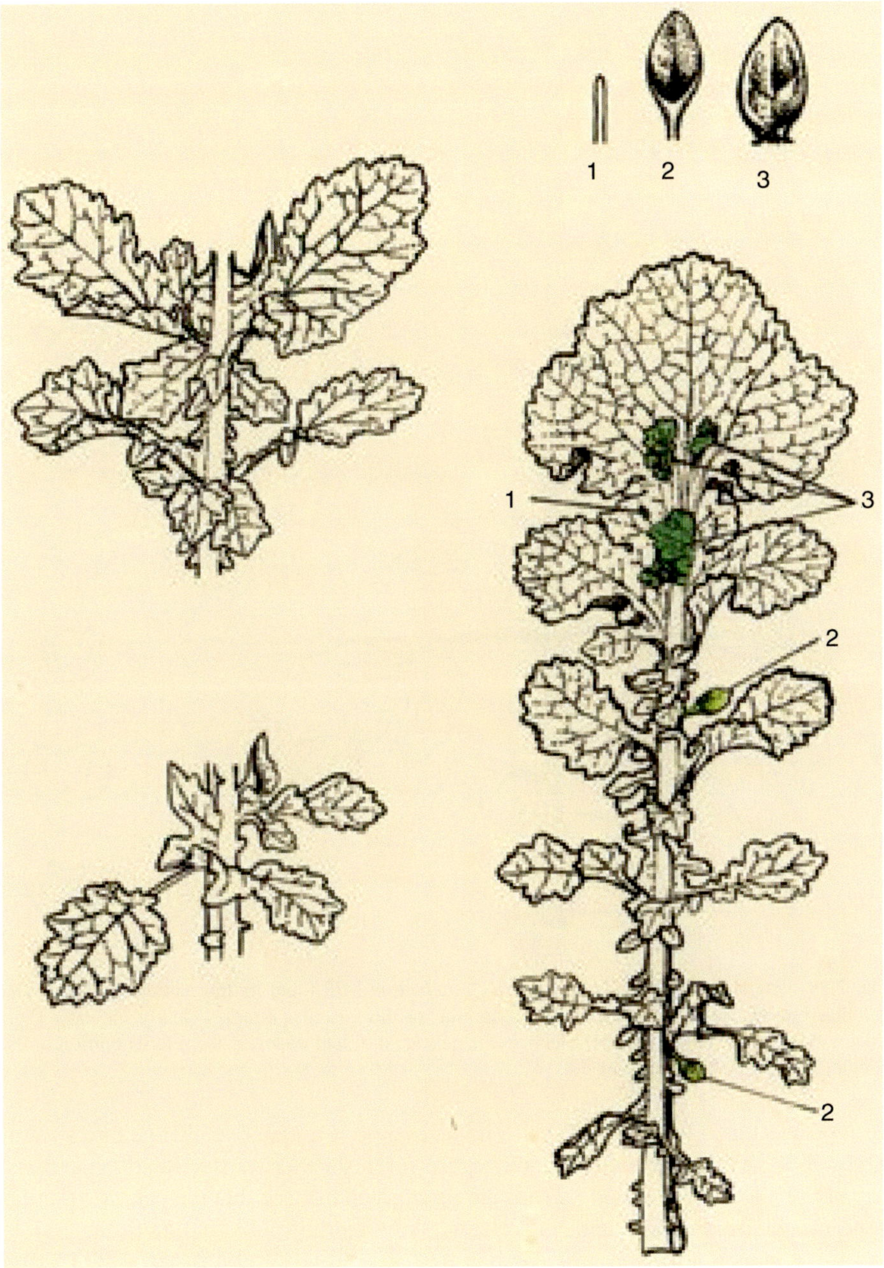

Fig. 7.38 The vascular bundle system of the phyllomes of mustard, showing the different stages of a newly formed axis flattening into a leaf. (1) Column-like unsplit newborn axis; (2) a horse-ear-shaped axis with a split upper portion; (3) a fully flattened leaflet

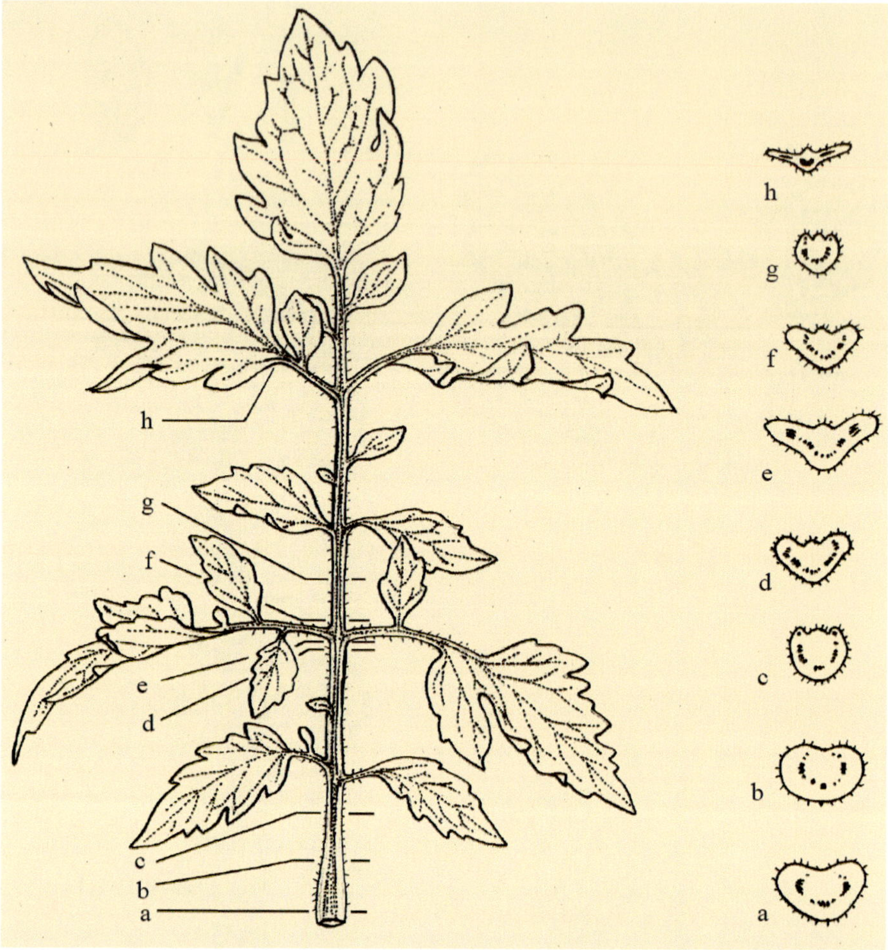

Fig. 7.39 The leaf of tomato (*Lycopersicon esculentum* Mill.) and its transections, showing the vascular bundle system (the half-ring distribution of the vascular bundles clearly indicates their axial nature); a: the base of the petiole, b, c: the petiole, d–f: leaf nodes, g: the middle portion of the petiole, h: the base of the leaflet

turn, proves that a leaf is a split axis and its petiole is not able to change into a stem, although it can grow branches again. Figure 7.41a shows a transection of an axillary bug at the petiole leaflet node. From the figure, you can see that the petiole vascular bundles are arranged as a half ring, clearly showing the split structure of the petiole and the connection of the vascular bundles of the bud grown in the leaflet axilla (indicated by the arrowhead in Fig. 7.41a, b) to the petiole vascular bundles. The axillary bugs have a standard sympodial stem structure, and petioles have a standard split axis structure.

Peach (*Amygdalus persica* L.) is a famous ancient fruit and flower tree native to Asia. The oldest book in China, "The Book of Songs," recorded these poems about peach: "The peach tree is elegant, brilliant when it blossoms; a young lady is going to

Fig. 7.40 Axillary bugs from leaflet axillae of tomato's compound leaves and the branches developed from them

leaves, to her future home she well suits" and "I was given a peach, I give back a jade; not return it for thanks, but for a friendship that lasts." Peach flower is ovary-perigynous. Its calyx, petals, and stamens all grow on the calyx tube, a specially deformed structure surrounding the ovary. These organs have evolved from dichotomous axes. Both the calyx and the petal are phyllomes formed from leaf clefts. The

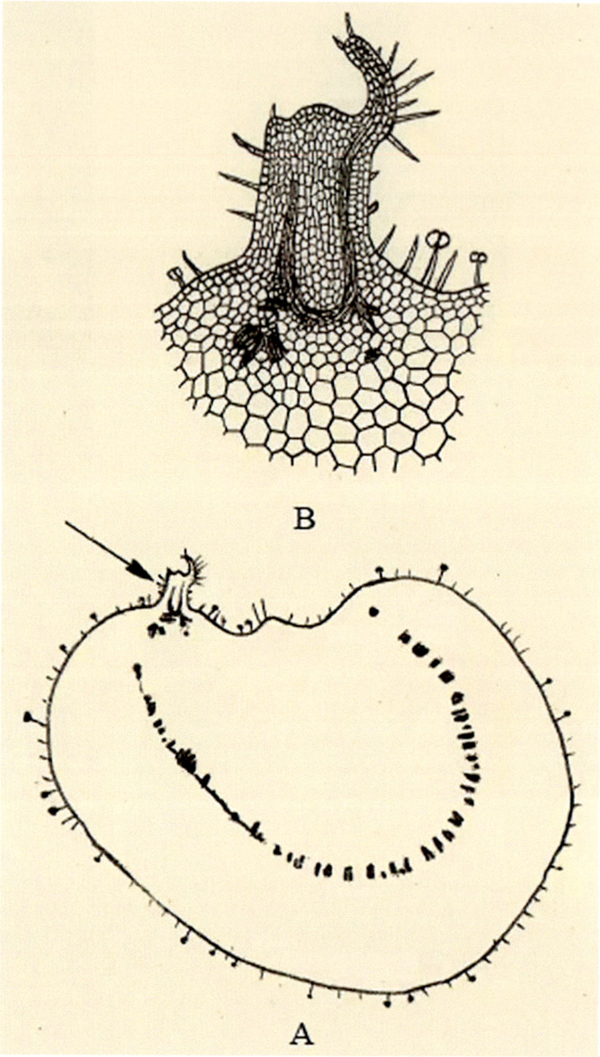

Fig. 7.41 The transections of a tomato petiole and an axillary bug. (**a**) The transection of a petiole, showing an axillary bug (arrow head); (**b**) the enlargement of the axillary bug

upper portion of the dichotomous axis develops unevenly. One side rises, leading to its dislocation and thus forming an alternate phyllotaxy. Consecutive dislocations of the secondary axes cause a difference in the area, resulting in phyllotactic rotation. Five rotating leaf clefts are aggregated together to form the pentaphyllum sepal and the pentaphyllum corolla. The lower portions of the sepal and corolla are aggregated and then grow downward to form the calyx tube. Stamens grown on the upper end of the calyx tube are evolved from the axillary buds of the petals and calyx. It should be noticed that the vascular bundles of the calyx, i.e., the veins of this phyllome, have five vascular bundles that extend downward, the two at each side merging with the vascular bundles of two petals, respectively. We can also say that three vascular

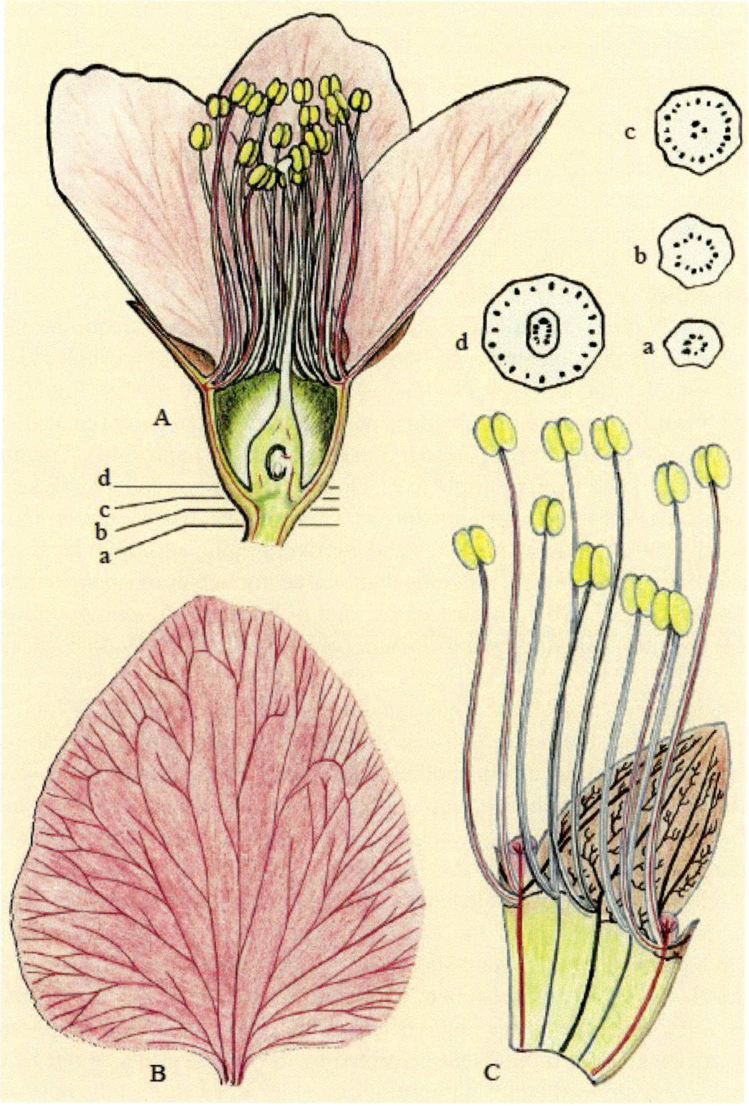

Fig. 7.42 The vascular bundles of the flower of peach (*Amygdalus persica* L.). (**a**) The longitudinal section of the flower and four transections, a–d; (**b**) petal vascular bundles, showing the merging of the upper portion of the three systems; (**c**) the vascular bundle system of the calyx tube, sepal, and stamen; the lower portions of the three vascular bundles are not interconnected but are connected with petiole vascular bundles, respectively

bundles of the petals extend downward and merge with the two vascular bundles from the calyx to form five of the ten main veins underneath the petals. The other five come from the calyx alone. Ten smaller vascular bundles also come from the calyx. The 20 vascular bundles in the calyx tube are also connected with the thin filigree vascular bundles of the two axillary stamens of their own (Fig. 7.42). One calyx contains five relatively independent vascular bundles. If we follow the telome theory and regard a leaf vein as a branch, we could not explain this constitution. It is even harder to explain that two of the vascular bundles underneath the petal lead to the calyx, respectively. According to experimental biology, vascular bundles have been proven to be the product of secondary development induced by growing hormones, which clears the fog fabricated by "the telome theory." The vascular bundles in the calyx tube are induced by the auxin flows down from the calyx, petals, and stamens, which together promote the differentiation of the 20 vascular bundles in the calyx tube. At the beginning of the calyx tube formation, there are only parenchyma but no vascular bundles. Vascular bundles are secondary.

Plane tree (*Platanus* L.) is also a plant with very special constructions. There are three species, and they are basically similar to each other in morphological structures. The leaf of a plane tree looks like a palmate leaf at first glance (Fig. 7.43). If it is viewed from its anatomical sections, its structure is very complicated, and its phyllotaxy is also different from ordinary leaves (Figs. 7.44 and 7.45). The nodes on the stem have evolved from multiple axes. On the basis of structural analysis, a stem node is composed of three overlapped axes. This scenario has been seen in wheat's node, which is a compound node formed by overlapping multiple axes. Multiple axes are formed on the lower leaf clefts and evolve into a pair of opposite palmately simple leaves (Fig. 7.44b). In most cases, one of the opposite compound palmate leaves degenerates and disappears, leaving only one leaf to become an alternate phyllotaxy (Fig. 7.44a). Occasionally, one of the opposite leaves is slightly dislocated and thus rises, but it has not yet become alternate (Fig. 7.44c, d). In Fig. 7.44c, d in the internode, you can see additional stippllum above the stipule, which indicates that there exists another axis, which also indicates that this node is composed of at least three overlapping layers of axes. Its alternate phyllotaxy is formed by the degeneration of one of the opposite leaves, and the specimen shown in Fig. 7.46a has confirmed this. In this specimen, a nonopposite simple leaf lies at the node, which should have opposite leaves, and the upper node has another opposite. If the simple leaf is like one of the ordinary opposite leaves, which are dislocated upward to become alternate, there should be another simple leaf node located above this node. However, in this specimen, above the simple node, there is another node of opposite leaves, too, and the angle between the opposite and the simple nodes is 90°. This means that the simple leaf is formed by the degeneration of one of the two leaf clefts, which are formed from the split axes of the third axis on the opposite axis. Figure 7.46b also shows that the opposite leaves are offset by 90° but not by 180°, or the opposite leaves rise spirally more than 180°. This is also evidence.

The leaves of plane trees, as discussed above, have a very complex structure. From Figs. 7.45 and 7.47, we can see that three major veins merge downward into the main veins to become a layer, and three layers of up to nine vascular bundle systems go down. Obviously, it is a secondary constitution, with each component being developed from its own axes, respectively. This portion of the main vein obviously is a composite tissue constituted by intermerged multiple axes. It is not a branch surmised by the telome theory nor a simple leaf assumed by the phytonic

Fig. 7.43 Leaves of plane tree (*Platanus* × *acerifolia* (Ait.) Willd.) in different stages of development. (**a**) A newborn leaf, (**b**) a further developed newborn leaf, (**c**) a palmate leaf just in shape; (**d**) a further developed palmate leaf; (**e**) a fully developed palmate leaf

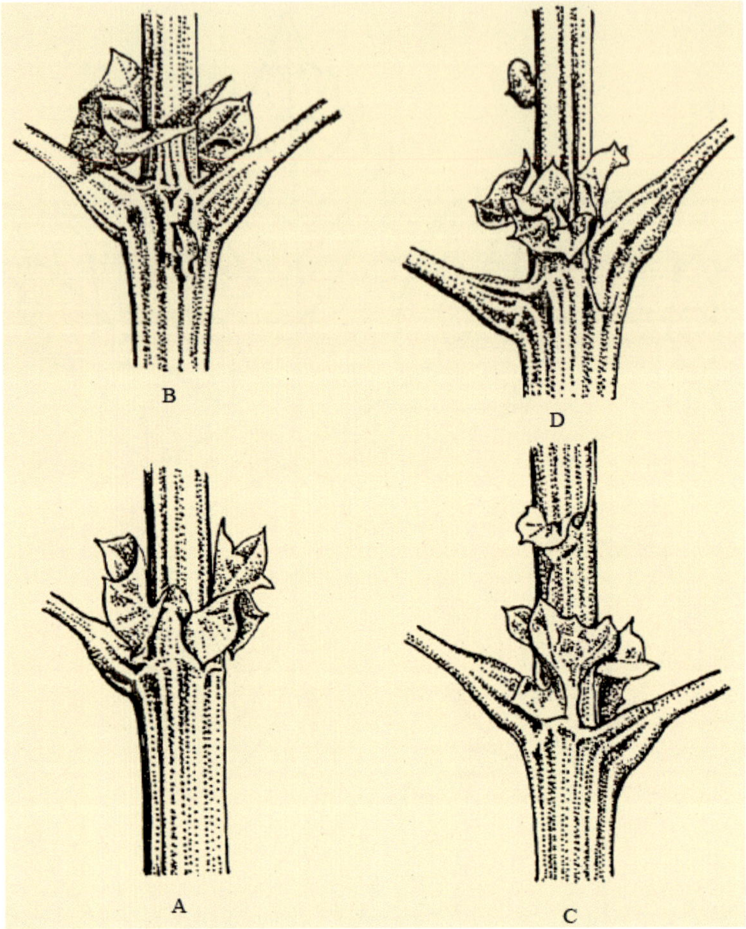

Fig. 7.44 The phyllotaxy of London plane tree (*Platanus* × *acerifolia* (Ait.) Willd.). (**a**) An alternative arrangement resulting from the degeneration of one of the opposite leaves; (**b**) normal opposite phyllotaxy; (**c**, **d**) the views from both sides of the growth of misaligned opposite leaves, please pay attention to the cauline stipllum above them, which indicates the multiple structure of the stem

theory. Instead, it is a compound leaf constructed by splitting and then intermerging (webbing) multiple axes.

Opposite leaves can be seen on paper mulberry (*Broussonetia papyrifera* (L.) L'Hér. ex Vent.) of the Moraceae family; the degeneration of one leaf has made it smaller (Fig. 7.48). However, whether one of its two opposite leaves has disappeared to become an alternate phyllotaxy, as is seen in plane tree, the author has not yet seen any direct evidence of it. Nevertheless, apparently similar alternate phyllotaxy could be formed by single splitting in the upper portion of the axis with a realignment angle of 180° between, just like the monocotyledons of Gramineae, or by degeneration of one of the opposite leaves developed by splitting the upper portion of the axis with

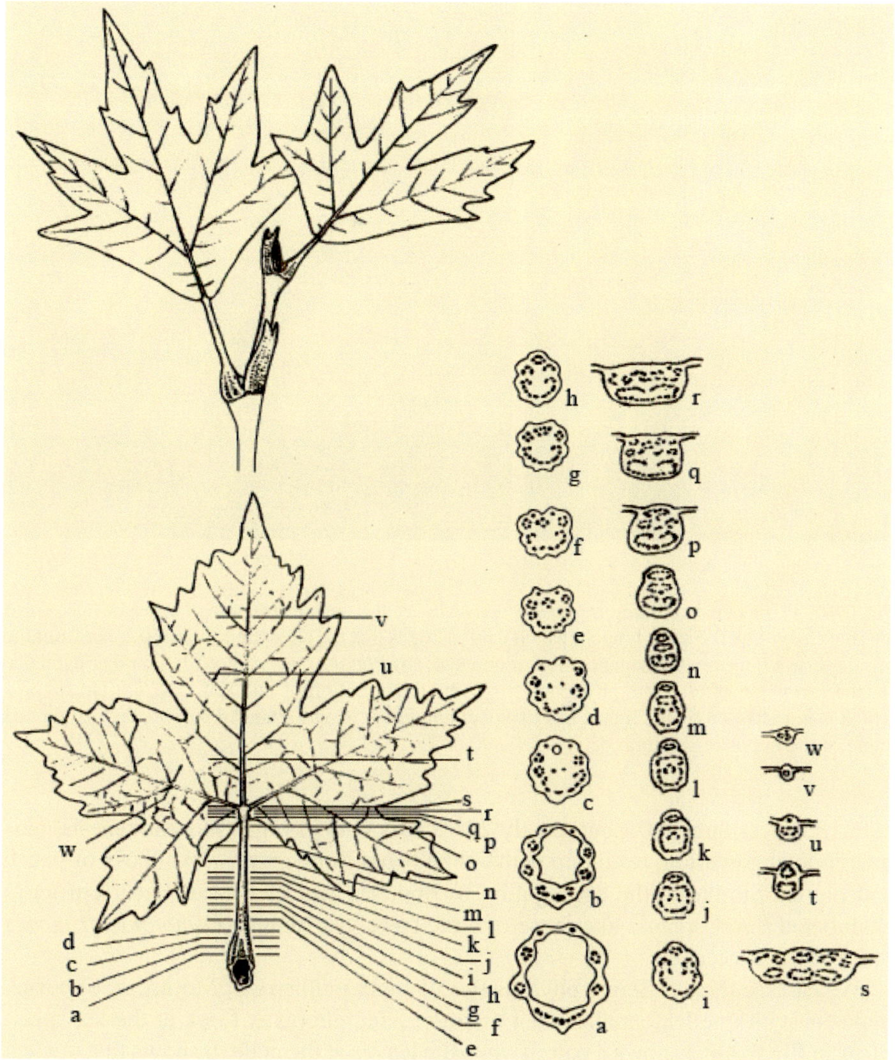

Fig. 7.45 Transections of different parts of the leaves of London plane tree (*Platanus* × *acerifolia* (Ait.) Willd.), showing an overlapped development of vascular bundles. (The typical alternate phyllotaxy at the top obviously results from the degeneration of one of the opposite leaves. The opposite bracts are the remnants)

an angle of 90°, like those of plane tree, or by an imbalanced development between two opposite leaves in a way that the growth of one leaf in a pair lags behand the other so that the fast growing one rises up its position as the composite stem grows, resulting in the apparent alternate leaves, like those of ginkgo, peach, plum and paper mulberry. Such an imbalanced development causes one leaf to rise, which often causes an area difference and thus the rotation of the phyllotaxy to an angle of greater

A B

Fig. 7.46 Plane tree. (**a**) *Platanus* × *acerifolia* (Ait.) Willd., with alternation of opposite leaves and alternate leaves on the same branch, which demonstrates that the alternate leaves result not from the misalignment but from the degeneration of one leaf; (**b**) *Platanus occidentalis* L., another evidence for the formation of alternate phyllotaxy from the degeneration of one leaf. The upright branch, which is not affected by the lights, has its upper and lower leaves arranged by 90°, the maintained angle of the opposite phyllotaxy

than 180°, less than 360°, but mostly 298°. This kind of organ morphogenesis, plus internode aggregation, results in penta-chambered floral organs, like those of peach and plums. Similarly, the aggregation of opposite leaf clefts can also form tetra-chambered floral organs, like those of Cruciferae, osmanthus (*Osmanthus fragrans* Lour.) of Oleaceae, etc.

We can see three different phyllotaxies in paper mulberry of Moraceae: opposite, opposite with one leaf smaller than the other, and alternate. Look at the six specimens in Fig. 7.48: a shows a pair of opposite leaves at the node; b shows that the leaf cleft at one side rises to become an alternate phyllotaxy; c shows that the phyllotaxies on the opposite branches, which grow in the axillae of the first-year opposite leaves, are alternative; d shows a branch with opposite leaves; e shows that one leaf of the opposite phyllotaxy on an opposite foliage branch degenerates to become smaller; f shows that, on an alternate foliage branch, the upper and lower leaves in a group have an angle of 180°, but the upper group, which has an angle of 180° between the leaves within the group, has a 90° angle to the lower group. Here, we can see, on the same paper mulberry tree, that opposite and alternative phyllotaxies can coexist, that a branch can simultaneously have opposite and alternative foliar nodes on the same branch, and that phyllotaxies grow on the same branch but can be opposite or alternative in a different year. It seems that, in paper mulberry, whether a

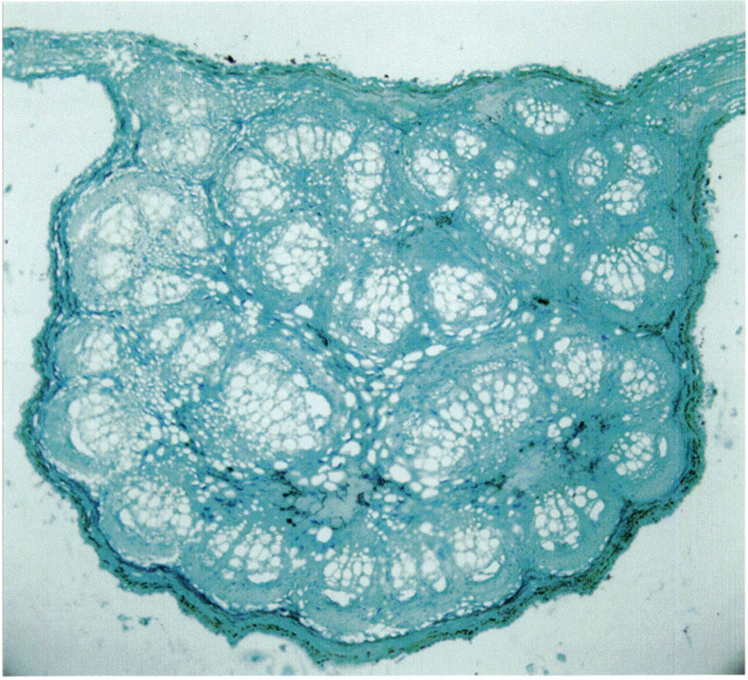

Fig. 7.47 The vascular bundle structure of the main vein of plane tree, showing the transection of the three layers of a total of nine vascular bundles. The location is equivalent to that shown in Fig. 7.44p, q. The slide was made by Ji Liu and Hongyu Yang

phyllotaxy is opposite or alternative is determined by what kind of leaf cleft is induced by the external environmental conditions during the growing season. A lag in the growth of one leaf cleft in an opposite pair will cause the rise of the other side of the nose, which leads to the change from an opposite phyllotaxy to an apparently alternate phyllotaxy. Therefore, this change is not controlled by internal genetic factors alone but results from the genetic interaction induced by external factors.

From the case of plane tree, we can see a phenomenon: dicotyledonous opposite leaves on the upper and lower nodes have an angle of 90°. If one of the opposite leaves degenerates, the leaves on the upper and lower nodes will still have an angle of 90° (Fig. 7.48). However, if the leaves (opposite or one leaf degenerate) on the upper and lower nodes grow on a horizontal branch, they may, induced by the light from the side of the branch, flatten out toward the light to an angle of near 180° (Fig. 7.48a, b). This effect is caused by auxin's apheliotrophism, which causes the cells at the shady side of the petiole and branch to elongate and deform. This mechanism is just as that of the anthodium of sunflower (*Helianthus annuus* L.), which always turns toward the sun. If the light is from the top of the branch, light-induced auxin will not accumulate, and the angle between the upper and lower leaves will remain 90° (Fig. 7.46).

Fig. 7.48 The phyllotaxy of paper mulberry (*Broussonetia papyrifera* (L.) L'Hér. ex Vent.). (**a**) Equally grown opposite leaf clefts; (**b**) unequally grown leaf clefts with one rising to become alternative; (**c**) a 2-year-old branch with opposite phyllotaxy for the first year; the new branches developed from the axillary buds are all opposite but have alternate phyllotaxy on them; (**d**) an opposite foliage branch; (**e**) an opposite foliage branch; (**f**) an alternate foliage branch, which is well recognized as the standard morphological structure of paper mulberry

The leaves of the aforementioned paper mulberry of Moraceae, plane tree of Platanaceae, peach and plums of Rosaceae were all regarded as simple leaves in old botanical books. In fact, they are all compound phyllomes formed by splitting, flattening, webbing, and evolving of multiple secondary axes. Because it is a compound system composed of multiple axes, its secondary vascular bundle system also reflects its complex construction. The leaf of plane tree, shown in Fig. 7.47, is such a typical construction. It is completely different from the simple leaf construction of lycopod, spike moss, cypress, cedar, and ephedra. It is formed not from the leaf clefts of one axis but from multiple secondary axes through aggregation and webbing. This kind of leaves formed from aggregation and webbing of multiple secondary axes need the adjective "composite" to describe or express it. The leaves of wheat, barley, onion, garlic, cabbage, mustard, plane tree, paper mulberry, etc., which were called simple leaves, should be called composite simple leaves. Having composite simple leaves is a major feature of angiosperms. The leaves of Pteridophyta and Gymnosperm formed from multiple axes are all composite leaves. The aforementioned *Helminthostachys zeylanica*, *Angiopteris*, *Pteris serrulata*, *Dicranopteris dichotoma*, and *Lygodium japonicum* of Pteridophyta and *Cycas* of

Gymnosperm are examples. Simple leaves are all composed of a single axis, such as those of *Psilotum* of Psilopsida; those of *Azolla imbricate*, *Ophioglossum nudicaule*, Selaginellales, lycopodia, and Sphenopsida of Filicopsida; and those of Pinaceae, Cupressaceae, Ginkgoaceae, Ephedraceae, and Welwitschia.

One is advised to pay attention to this: the oppositeness or alternateness represented by the opposite composite simple leaves or the alternate composite simple leaves of plane tree and paper mulberry are the oppositeness or alternateness of the stipules. This is because these composite simple leaves are those formed from the secondary axes that grow in the axillae of the stipules of the leaf clefts of the first axis.

References

Ikeno S (1906) Plant systematics (Chinese translation by Luo Zongluo published in 1947 published by The Commercial Press, Shanghai)

Kihara H (1937) Genomanalyse bei *Triticum* und *Aegilops*. VII. Mem Coll Agr Kyoto Univ 41:1–61

Percival J (1921) The wheat plant: a monograph. Duckworth and Co, London

Chapter 8
Organogenesis of Higher Plants

The organs of higher plants are specialized organs with special characteristics that have evolved through a series of physiological and biochemical processes directed by the plant's own genetic materials under particular environmental conditions. Thus, these specific traits are also specific genetic traits. Many people believe that specific heredity determines a particular trait, which, thus, is unchangeable. This view is actually wrong. In fact, the same inheritability can only express the same trait under the same environmental and physiological conditions; it will express different traits if the environmental and physiological conditions are different. Different heredities can express the same trait under different environmental and physiological conditions, but they can also express different traits under the same environmental and physiological conditions.

Taking common wheat (*Triticum aestivum* L.) as an example, starting from the fertilization of the egg in the embryonic sac by the sperm from the pollen tube, it has paired chromosomes from both parents and contains three subgenomes (ABD) and the whole set of paired deoxyribonucleic acid (DNA), the genetic materials. This set of paired DNA genetic materials is unchanged until the individual dies. However, because the external environments and internal physiological conditions are different, different genetic traits are expressed at different stages of individual development. As shown in Fig. 7.20, the morphology and structure of leaves of the same plant (the same individual) in different periods are completely different. The coleoptile consists of a single axis; the basal leaves are composed of multiple axes with a very short slit at the upper end of the sheath, the ligula and auricle are not developed, and the blade has no constriction; on the lower cauline leaves (the leaves on the main stalk during the tillering stage), the sheath splits incompletely, the ligula and auricle begin to form, and a constriction appears in the middle to the upper portion of the blade; on the middle and upper cauline leaves (the leaves on the main stalk after jointing), the sheath still does not split completely, and the ligula and auricle are well developed; the flag leaf has a completely split sheath, a well-developed ligula and auricle, and a constriction is located near the blade tip. They vary greatly in length, width, thickness, and size, although they have exactly the same genetic basis. When

C. Yen, *Constitution of Organs of the Higher Plants*,
https://doi.org/10.1007/978-981-19-1685-4_8

they grow and develop, the external environments and the internal physiological conditions are completely different; therefore, the expression of the same heredities are completely different because of the completely different external environments and internal physiological conditions. The morphological structure of the basal leaves is the genetic trait that is expressed during the foliate development before the completion of physiological changes by vernalization. The cauline and flag leaves develop under long daylength and warm conditions, above 5 °C after the completion of physiological changes caused by vernalization. If a wheat plant is grown for a period of time (the time required depends upon the heredity of a cultivar) under 5 °C and thus has completed the physiological changes induced by vernalization and then grown in long daylength and warm condition, it will undergo jointing, and its leaf clefts, therefore, present the expression of cauline leaves. Therefore, the heredity is the same, the external environment and internal physiological conditions are different, and the expression will be different.

On the same plant and under the same external environmental conditions, the expression of the heredity between different tillers may also be different. For example, if a spike-branching wheat plant is grown in a good open field, although the living conditions are good, and a majority of its tillers do not degenerate and thus develop spikes, the expression of the genetic trait of spike branching is still different between different tillers. The heredity of spike branching is expressed on the main spike, and the expression on the tillers is decreased in a descending order based on their tillering time. The spike of most upper (the latest) tiller may not branch to become an unbranched simple spike.

Cultivars that do not have a spike-branching trait under a normal growing condition can also produce branching spikes if being treated with short-day length by shading after the completion of the physiological changes induced by vernalization. In 1960, the author saw the results of a short-daylength treatment on wheat at the Indian Agricultural Research Institute in New Delhi, and among them was a spike-branching specimen.

A visitor asked a research staff: "Is this branching spike inheritable?" The staff answered: "No, it is not."

Everyone at that time understood what happened, and thus no more question was asked. However, it reflects that the questioner and the answerer had a wrong understanding of heredity and expression. If the questioner changes the question to, "Is the spike-branching spike inheritable without shading?" the understanding of inheritance by the questioner and the responder would both be correct because experiments have demonstrated that all wheat cultivars have the heredity of branching spikes. The only difference is that each requires different daylengths to express it. Those cultivars that grow branching spikes due to short daylength have the heredity of the branching spike trait. The expression of their branching spike trait by their DNA needs special short daylength. Therefore, for those cultivars that are called branching wheat, the daylength in their normal cultivating conditions are the special environmental condition under which their DNA expresses the branching spike trait. It is not that they have the genetic trait of a branching spike or not but that their DNA needs different daylengths to express the branching spike trait. Not only external environmental conditions but also internal physiological conditions will cause the same individual to express different genetic traits at different parts of the body or at different times (e.g., different ages). For instance, wheat can have a branching main spike and unbranched spikes on the tillers.

In the warm winter of Sichuan, Odessa 3, which has a strong winterness cannot complete vernalization and, thus, always remains in the tillering stage. The development of its multiple axes is always in the morphological construction similar to those of basal leaves (Fig. 7.20c). Ligules and auricles are undeveloped; the leaf blades have no constriction; the leaf sheaths only have a small slit in the upper portion, and it will not split to open until the growth of the axillary buds. If the same newly germinated Odessa 3 kernels are artificially vernalized (at 5 °C for 20 days), it can be adequately planted at the same time as Spring wheat cultivar Chengdu Guangtou (i.e., Chinese Spring) and joint and head the same way, growing four to six basal leaves and then four to six carline leaves with a semi-split sheath on the lowest cauline leaf and a fully split sheath on the flag leaf. Ligules, auricles, and constrictions on the cauline leaves are all well developed. Besides, 17–24 spike nodes with degenerated leaf clefts are formed through the aggregation of paired axes. The leaf clefts of the axis, which grow in the axillae of a ring of leaf cleft residues on the spike node, resume growing. Their upper portions develop into glume, lemma, and palea. The axillary axes of the lemma and palea then further develop into stamens and the pistil. All the axes are the same, growing bipolarly and having a structure at the upper ends that can expand and split. They, under different external and internal conditions, are constructed into organs of different shapes and functions on the wheat plant.

DNA in *Triticum* genomes (A^uA^u, A^mA^m, $B^{sp}B^{sp}$, $A^uA^uB^{sp}B^{sp}$, $A^mA^mB^{sp}B^{sp}$, $A^uA^uA^mA^m$ $B^{sp}B^{sp}$, and A^uA^uBBDD) undergo mutations. Those mutations that have survived the long-term natural selection gain adequate fitness, and they will express when the plants are grown under the intertwined influence from the external and the internal conditions. This indicates that an individual's genome does not change during its lifetime. It, under different external and internal conditions at different times, directs the construction of organs of different shapes and functions and the completion of the growth and development of the organs that are interconnected and interdependent to constitute different developmental stages of life. Figures 8.1 and 8.2 illustrate the longitudinal sections and growth cone of a wheat plant at different morphological developmental stages. They show that axes constitute the phyllomes of different shapes at different developmental stages: embryo leaf, coleoptile, basal leaf, cauline leaf, flag leaf, rachis ring, glume, lemma, palea, stamen, pistil carpel, and outer and inner integuments.

Figure 8.3 shows an abnormal stamen with an uninhibited growth cone. It demonstrates that an anther is composed of two leaf clefts, that a stamen is an axis, and that an axis can be formed from a leaf cleft's axilla and then develop into a growth cone.

Forming an axis needs not only a single cell but a cluster of meristematic cells. Usually, cells in this cell cluster contain the same genes and are all derived from the same fertilized egg. Usually, genes in these cells are the same since they are all

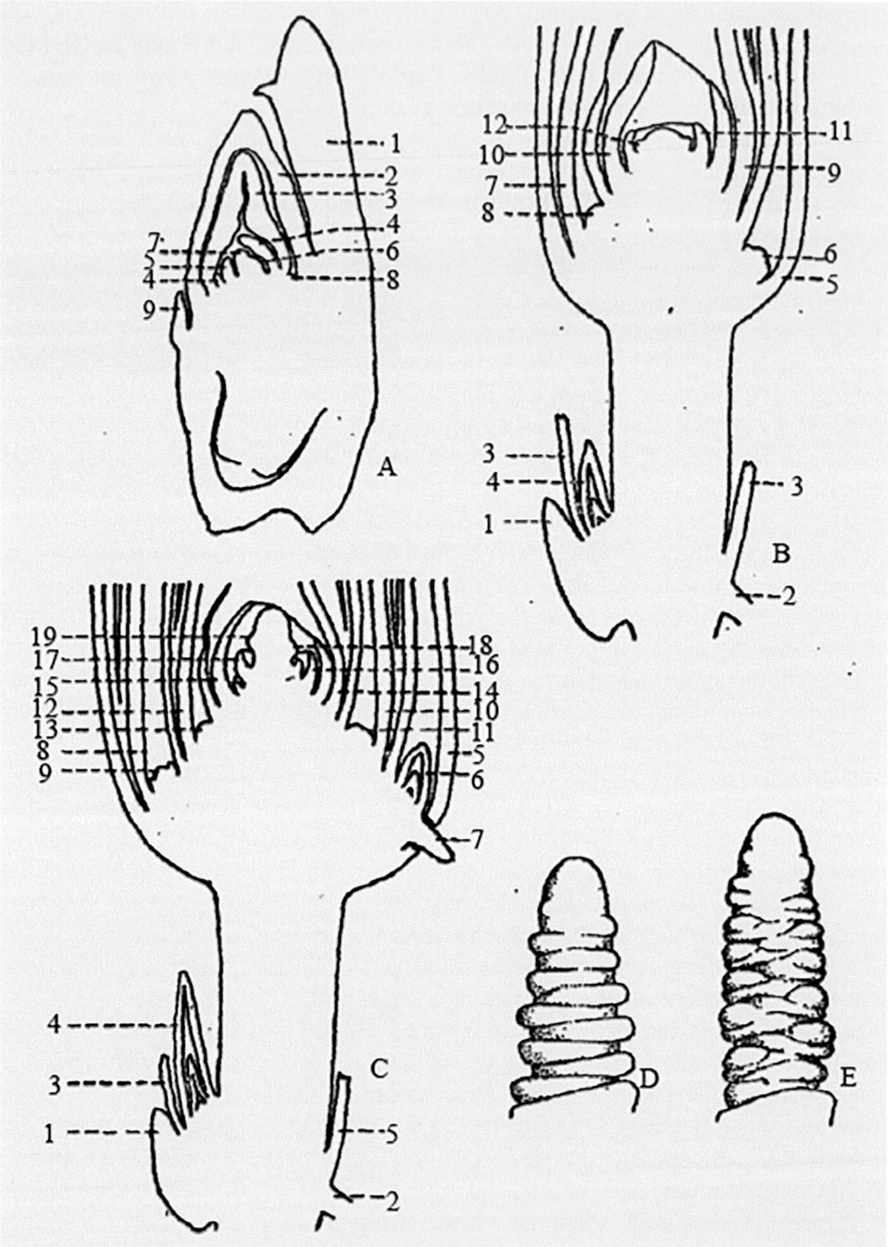

Fig. 8.1 The growth cone of *Triticum aestivum* in the first, second, and third developmental stages. (**a**) The longitudinal view of the embryo, showing the early first developmental stage; 1: scutellum, 2: coleoptile, 3: the first green leaf, 4: the second green leaf, 5: the primordium of the third green leaf, 6: the primordium of the fourth green leaf, 7: the growing point, 8: the axillary bud of the coleoptile (the first tiller bud), 9: the outer integument. (**b**) The longitudinal view of the seedling in the middle of the first developmental stage; 1: scutellum, 2: the seminal root opposite to the scutellum, 3: residue of coleoptile, 4: the axillary bud of the coleoptile (the first tiller bud), 5: the first green leaf, 6: the second tiller bud, 7: the second green leaf, 8: the third tiller bud, 9: the third

derived from the same fertilized egg. However, it is possible that mutation may occur in some of them, making this cell cluster a genetic chimera. This may cause the expression of different heredities in tissues of different parts of the different organs formed by the same axis. The most common is pigmentation chimera. For instance, spider plant (*Chlorophytum comosum* (Thunb.) Jacques), a South African native often used for bonsai, has a variety that has yellow-white stripe mosaic leaves due to chlorophyll deficiency. An axis can develop from a cluster of inosculated meristems derived from grafted plant parts of different heredities, and it is, therefore, not uncommon, as many have been recorded, that different parts of an organ may express different heredities. These examples all demonstrate that an axis develops from a cluster of meristematic cells.

Let us also look at a few cruciferous species of *Brassica* L. Let us look at kale (*B. oleracea* L.), the type species of the genus first. It is a common annual/biannual plant in England and the subtropical Mediterranean seashore, growing in autumn and winter and blooming and setting seeds in spring and summer. People like to eat its stems, leaves, and flower buds and have had it cultivated since ancient times. In kale's whole life, heredity does not change, but its expression changes as internal physiological conditions and the external environment change, forming multiaxial organs with various shapes. The nascent euphyllae have a long petiole. This long petiole has an egg-shaped blade on the top and one to three small leaf clefts somewhat connected to the lower end of the blade. The leaves on the flower stalk do not have petiole and the leaf shapes are from oblong to lanceolate. The difference between its calyx, petal, stamen, and pistil is even greater. They all have the same heredity, but their expression varies when internal physiological conditions and the external environment change, forming phyllomes of various shapes composed of one or multiple axes.

On the other hand, a mutant cultivar with a prolonged first morphological developmental stage has more long-petioled leaves. At this stage, the stems composed of compound axes develop lots of parenchyma cells in the pith of the xylem, forming a swollen tender corm. People like this abnormal stem. Therefore, this genetic variation has been preserved to become a special true-breeding cultivar named kohlrabi (*B. oleracea* L. var. *gongylodes* L.). Except for corm, kohlrabi and kale have little difference in morphological traits, and there is no reproduction barrier between them either.

Fig. 8.1 (continued) green leaf, 10: the fourth green leaf, 11: the primordium of the fifth green leaf, 12: the primordium of the sixth green leaf. (**c**) The longitudinal view of the seedling in the early second developmental stage; 1: scutellum, 2: the seminal root opposite to the scutellum, 3: residue of coleoptile, 4: the first tiller bud, 5: the first green leaf, 6: the second tiller bud, 7: the first tiller root, 8: the second green leaf, 9: the third tiller bud, 10: the third green leaf, 11: the fourth tiller bud, 12: the fourth green leaf, 13: the fifth tiller bud, 14: the fifth green leaf, 15: the sixth green leaf, 16: the seventh green leaf, 17: the primordium of the eighth green leaf, 18: the undeveloped primordium of the first rachis, 19: the undeveloped primordium of the second rachis. (**d**) The growth cone in the late second developmental stage. (**e**) The growth cone in the late third developmental stage. (From Figure 1 of Yen and Ngao 1963)

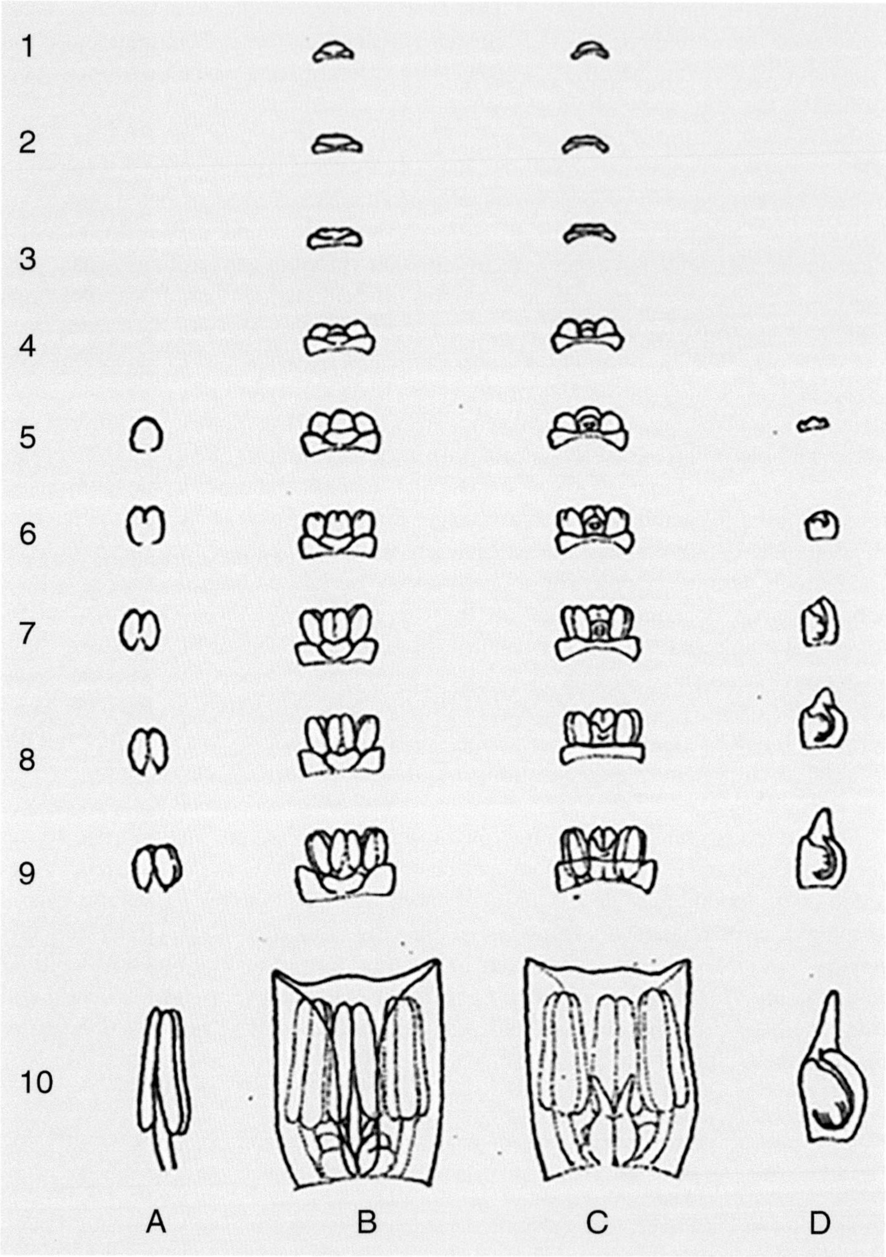

Fig. 8.2 Differentiation of wheat floral organs. (**a**) Stamen, (**b**) the ventral view of a floret, (**c**) the dorsal view of a floret, (**d**) pistil. (From Figure 3 of Yen and Ngao 1963)

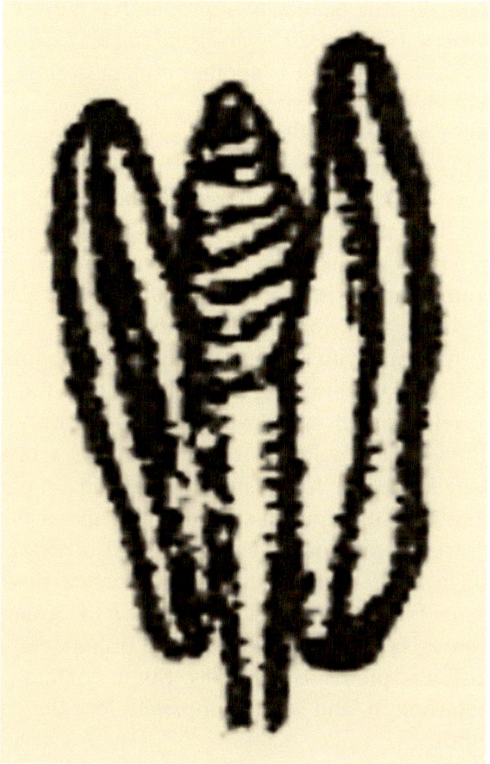

Fig. 8.3 A stamen with uninhibited apical growth, forming an apical cone (from Kuperman 1953)

A kale mutant with shortened first morphological developmental stage and prolonged second morphological developmental stage has a few long-petioled leaves but more petiole-less oblong leaves because the petiole-less leaves grow so fast that they stay very close to each other before the blades open up. As a result, they wrap each other to form a leaf ball. People like to eat the tender white inner leaves and thus have preserved this genetic variation and bred the special true-breeding variety named cabbage. C. Linne, the pioneer of scientific classification, correctly named this mutant *B. oleracea* L. var. *capitata* L. as a variant of kale. Cabbbage and kale have no reproductive barrier between them, and thus it is just right to treat it as a variant of kale. This kind of malformed leaf ball is not beneficial to the plant itself. It could not flower if it is not cut open by people, not mention seed setting. These mutants, which adversely affect the survival of the plant itself, would be eliminated by natural selection as soon as they appear because they could not set seeds.

Another mutant group, which is very similar to cabbage, also has a shortened first morphological developmental stage and a prolonged second stage. Unlike cabbage, their petiole-less rectangular leaves do not wrap each other to form a leaf ball. They are ornamental plants with dark purple or red anthocyanins in the upper inner leaves, which may also be yellowish white due to the degeneration of chlorophyll. They and kale do not have any reproductive barrier between them, and they are named *Brassica oleracea* L. var. *acephala* L. f. *tricolor* Hort..

Brussels sprout is another mutant of kale, which is also very similar to cabbage and also has a shortened first morphological developmental stage and a prolonged second stage. Unlike cabbage, its petiole-less rectangular leaves do not wrap each other to form a leaf ball. Just like kale, the internodes between the leaves elongate. Small leaf balls develop from the buds in the axillae. Brussels sprout and kale also do not have any reproductive barrier between them. Zenker named it *B. oleracea* L. var. *gemmifera* Zenker (i.e., Brussels sprout mutants) in 1922. Brussels sprout is different from cabbage in that the development of the distorted wrapping leaves is physiologically later than those of cabbage.

Mutants with shortened first and second morphological developmental stages but a prolonged third stage and the longest fourth stage enter the third morphological developmental stage when they have only one to two long-petioled leaves and almost no petiole-less oblong leaf. As a result, they develop about ten large and long lanceolate leaves, which are the main photosynthesis phyllome. The prolonged fourth developmental stage leads to the development of many flower buds on thick flower stems, forming a tender ball of flower buds without entering the flowering stage. This variant is, therefore, called flower-ball kale or cauliflower. C. Linne named it *B. oleracea* L. var. *botrytis* L. Cauliflower and kale have no reproductive barrier between them, Like the ornamental kale, this group of mutants also has different pigment mutations. Some contain purple anthocyanins, some have green chlorophyll, and some completely lose the color, becoming white, forming various cultivars.

This species has accumulated a large number of mutations under long-term cultivation, forming various varieties or, more precisely, concultivars due to artificial selection, such as the aforementioned kohlrabi (stem-ball kale), cabbage (leaf-ball kale), ornamental kale, Brussels sprouts, and cauliflower (flower-ball kale). Because of their large differences in morphology, they were incorrectly classified as species by some taxonomists, who at that time did not know their phylogenetic relationship. For instance, kohlrabi was named *Brassica caulorapa* (DC) Pasq.; Brussels sprouts, *Brassica gemmifera* Levl.; and cultivated kale, *Brassica alboglabra* L.H. Bailey. With a deeper understanding of their genetics, their phylogenetic relationships are now very clear. However, there still are some people who "retain the traditional concept to treat each of these vegetables as an independent species because the people have used them widely." This attitude is of course wrong for scientific research.

In *Brassica*, some other species, such as *Brassica chinensis* L. and *Broassica juncea* (L.) Czern et Coss., have completely similar organogenesis with that of *Brassica oleracea* L. From the phyllomes of Chinese cabbage, shown in Fig. 7.36, it can be seen that cotyledons, leaves with petioles, petiole-less leaves, calyxes, and petals are all axes containing the same genes. They form, on the same plant, different phyllomes of very different shapes under different external environmental and internal physiological conditions. This also demonstrates that the same genes can cause the formation of organs of different shapes because of the different changes in internal factors induced by different external factors. Gene mutation(s) cause

(s) minor modification(s) in response to external factors, resulting in change(s) in internal factor(s) and thus in different morphologies of the corresponding organ (s) among the variants.

Just like kale, the original types of Chinese cabbage, such as *Brassica chinensis* L. var. *oleifera* Makino et Nemoto (yellow rapeseeds) and *Brassica juncea* L. var. *gracilis* Tsen et Lee (high rapeseeds), have no obvious abnormality in their organs, except that phyllomes of different shapes are formed under different external environment and internal physiological conditions during their development. However, similar mutations occur in various developmental stages of these species, which cause abnormities. For example, *Brassica chinensis* L. also has a leaf-ball abnormality called Chinese cabbage (it is mistakenly named *Brassica pekinensis* (Lour.) Rupr.). Chinese cabbage and yellow rapeseeds do not have any reproductive barrier between them. Purple stalk, a variety of this species, has the same organogenetic mechanism as the ornamental kale of *Brassica oleracea*. *Brassica juncea* L. also has plants with swollen stem variation, which follows the same developmental mechanisms as kohlrabi. The corm mustard (*Brassica juncea* L. var. *megarrhiza* Tsen et Lee) has the same developmental mechanism as that of turnips of the same genus. It also has the same development mechanism as that of the fleshy root of radish (*Raphanus sativus* L.). Physiologically, corm mustard, turnip, and radish all have strong winterness, which is embodied in phyllomic morphology as a dense cluster of petiole leaves. After the completion of vernalization, these plants bolt and flower under long daylengths, developing petiole-less lanceolate leaves on pedicels. Before the completion of vernalization when daylength is not long, parenchyma cells in the xylem of the root actively divide to store photosynthetic products, resulting in an enlarged fleshy root. If the photosynthetic products are stored not in the root but in the stem, it becomes kohlrabi or succulent stem mustard (*Brassica juncea* L. var. *tsatsai* Mao). The photosynthetic products can also be used for the formation of tillers, lots of axillary buds in the first developmental stage, and clusters of petiolate leaves, and this is the exact developmental mechanism of leafy mustard (*Brassica juncea* L. var. *multiceps* Tsen et Lee). These morphological development mechanisms look varied a lot, but they orderly follow the logics. The discussions above are mainly about the organogenesis of a few abnormal roots, stems, leaves, and inflorescences.

The author published two papers in Bulletin of Botany in 1957 and 1958 to analyze how the silique and stamen of crucifers are formed. These two papers compared the intermediate forms of abnormal inflorescence and made it clear that "the silique of *Brassica* consists of two circles of carpels in the same section. Of the four sterile carpels on the outer circle, one pair constitutes the central part of the shell-shaped carpels, while the other pair forms the outer tissue and the lateral portion of the fertile carpels. One pair of the inner fertile carpels each combines ventrally with the midrib of one of the outer sterile carpels, respectively, forming a fruit carpel. The ovules are formed from the growth cone of the lateral branch formed on the inner fertile carpels, and fertile carpels on the inner whorl are not involved in the formation of style and stigma. The diaphragm, as a bridge structure, is formed along the dorsal longitudinal axis of the inner fertile carpels." At that time, the author did not know that "the multiple secondary axis" is the basic construction unit of higher plants. Although the structural relationships between the various parts of the silique were clarified, he, at that time, did not understand the fundamental origin of the tissues and their basic structural constitution. The malformed siliques are very clear on the rapeseed variety Victory (Fig. 8.4). A highly malformed silique shows a structure of two green leaves. Figure 8.4 and 8.5 clearly shows two rows of green leaflets along the main vein. The green leaflets along the main leaf vein of cauliflower were also correctly compared in Fig. 8.4(8), (9). These green leaflets are formed by splitting and flattening of multiple axes. Ovules also evolve from the regenerated multiple axes of those small green leaf blades. *Brassica*'s siliques are made from the two green leaves of multiple axes. Even the highly malformed silique (Fig. 8.4(3)) also exhibits the model of the two green-leaf-like structures, not that of four carpels. If a green leaf is recognized as a carpel, a silique is therefore composed of two carpels. This carpel is equivalent to a green rapeseed leaf. Not only does the green leaf produce many new axes to form leaflets along the main vein, but it itself is also formed from multiple axes. In addition, we must not forget that vascular bundles are secondary, and they are formed by transforming the parenchyma into xylem, phloem, and cambium, which are induced by the aggregation of growth hormones along their flow traces. It is incorrect that when you see a large vascular bundle, you regard it as one axis and it represents one carpel. The degeneration of a silique into two green leaves, as shown in Fig. 8.4(3), clearly demonstrates that a silique is composed of two carpels. The two carpels are large leaves composed of multiple axes, equivalent to a branch. The ovules are constructed by the axillary lateral buds formed on the multiaxial green leaf (carpel). Fundamentally, they are composed of multiple axes, which have nothing to do with the stigma construction at the top of the silique. The vascular bundle system we see in Figs. 8.5 and 8.6 has been seen in the leaf veins of plane tree (Fig. 7.48). It is an embodiment of the aggregation of multiple secondary vascular bundles differentiated from multiple secondary axial systems. In a plane tree, they come from the multiple leaf clefts of a compound leaf; in Victory rapeseed, they come from the silique shell and ovules. The formation of the outside main vein shown in Figs. 8.5 and 8.6 is induced by a growth hormone down from the stigma and style. Silique shell is formed by merging two carpels (each equivalent to a compound green leaf). The midrib is developed from parenchyma formed in the carpels under the inducement of growth hormones along the flow traces (Fig. 8.7). A malformed silique, as the one shown in Fig. 8.4(2) is formed by merging two green leaves, as those shown in Fig. 8.4(3). From their transections (Fig. 8.8), we can see, on the shuck's dorsal edge, a strong main vein lies at the middle of the seam formed by merging the edges of the two carpels but not at the middle of the carpels. This clearly indicates that vascular bundles are secondarily formed.

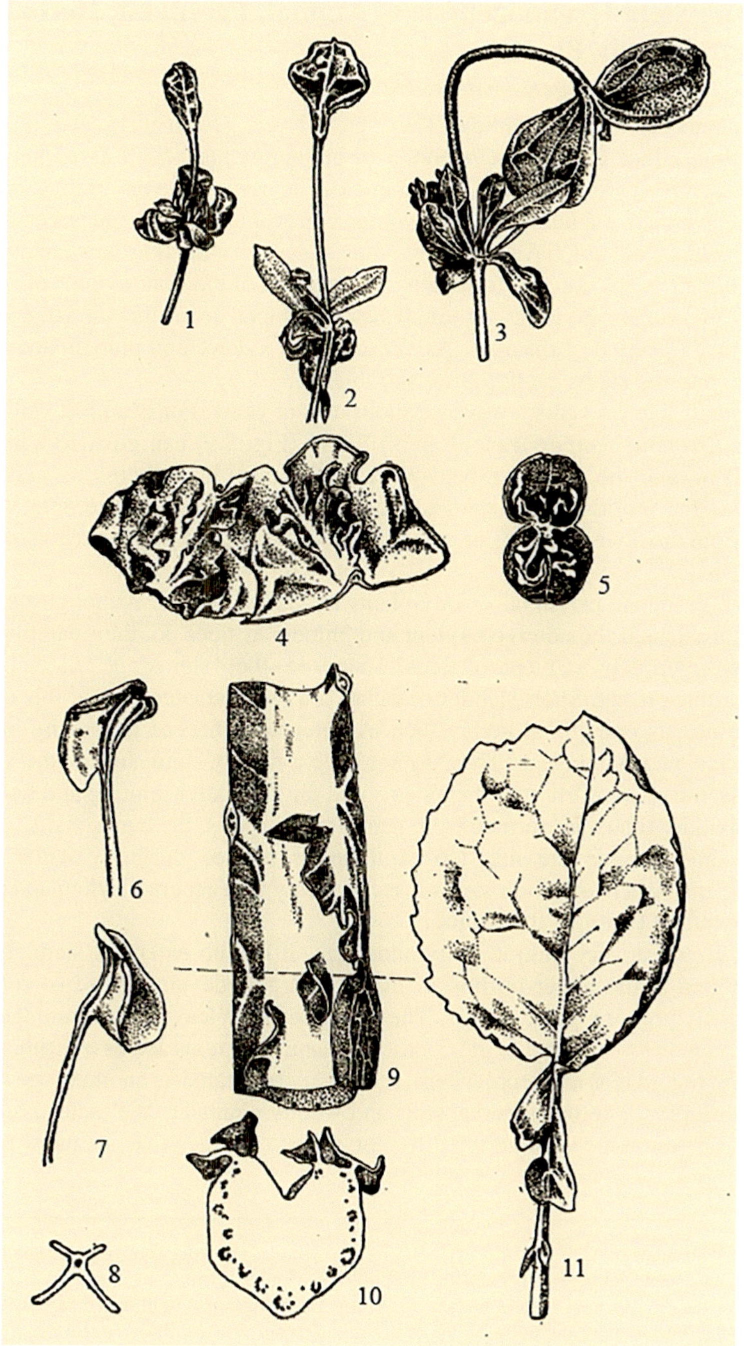

Fig. 8.4 Exterior morphology of a few malformed siliques of rapeseed cultivar Victory. (1) A silique with mild to moderate malformation; (2) a silique with moderate malformation; (3) a silique with high malformation; (4) a moderately malformed silique with the upper part cut to open and flatten, showing two rows of green leaflets along the main vein; (5) the transection of the upper

8.1 The Stem Organogenesis of Annual, Perennial, Herbal, and Woody Plants

Annual plants can also be biannual. They spend 1 year germinating seed, blooming, and fruiting, either from spring/summer to autumn/winter, such as spring wheat, hull-less barley, yard-long bean and pumpkin, or over the year from autumn to summer, so-called biannual, such as winter wheat, barley, and rapeseed. Annual/biannual plants do not have a strong stem of xylem and thus are called herbs. Although their stems are also formed by aggregation of the bottom ends of multiple axes, no vigorously dividing cambium has developed from the meristems in the merged lower portion of the axes. As a result, their xylems and phloem are not able to divide, and the stems cannot grow and thicken.

Perennials live for years, even over thousands of years. For example, Californian redwood (*Sequoia sempervirens* (Lamb.) Endl.) (Fig. 8.9) can grow to a height of over 100 m, and the stem can be 5–6 m in diameter. It is, therefore, called "the Grandpa of the World." Its tall stem is also representative of trees. The stem, which is formed from the lower portions of compound axes, is the most eye-catching organ of the plant.

Dwarf, cespitose perennial woody plants are called shrubs. Although they have similar cambia and thus derived xylem and phloem as trees do, their cambia do not divide as vigorously as those of trees. Therefore, their stems are relatively small without a thick trunk. Short slender branches are the characteristics of this organ.

There are also perennial herbs, such as onion of Liliaceae and many plants in Orchidaceae and Asteraceae. They do not have a woody stem. Just as the annuals/biannuals, the upper portion of their stems will die back after jointing and flowering, but the axillary tillers are formed from near-ground nodes, the corm or rhizome will survive and continue the life. No cambium is formed during vascular bundle development. As a result, their stems lose the ability to regenerate xylem and phloem and thus cannot continue thickening.

Welwitschia is very unique in gymnosperm. It has no extended stem. The leaf clefts formed at the upper portion of the second axis develop into two (or more) continuously growing green leaves. The axes above the leaves constitute the reproduction branches. This kind of stem structure cannot be regarded as a shrub, nor is it difficult to regard it as a perennial herb. It can only be treated as another type of stem.

The differences between herbaceous and woody plants are the major issue to be discussed below. This is about the different ways to construct the stems by merging

---

Fig. 8.4 (continued) portion of a moderate malformed silique, showing four rows of green leaflets curling in the seed chamber; (6) exterior morphology of a malformed rapeseed stamen; (7) the dorsal view of a malformed rapeseed stamen; (8) the transection of a malformed rapeseed stamen (anther part); (9) the exterior morphology of the middle vein of a cauliflower (*Brassica oleracea* var. *botrytis*) leaf blade with extra leaflets; (10) the transection of the middle vein at the dash line in (9); (11) a rapeseed leaf with leaflets curling back into an ear-like structure. (From Yen 1959)

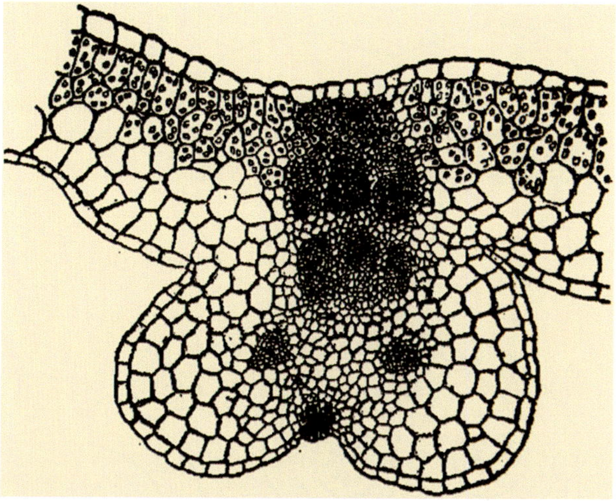

Fig. 8.5 The transection of a malformed rapeseed ovary wall, showing the morphology of the outer main fruiting vascular bundle, the inner main fruiting vascular bundle, and the surrounding tissues

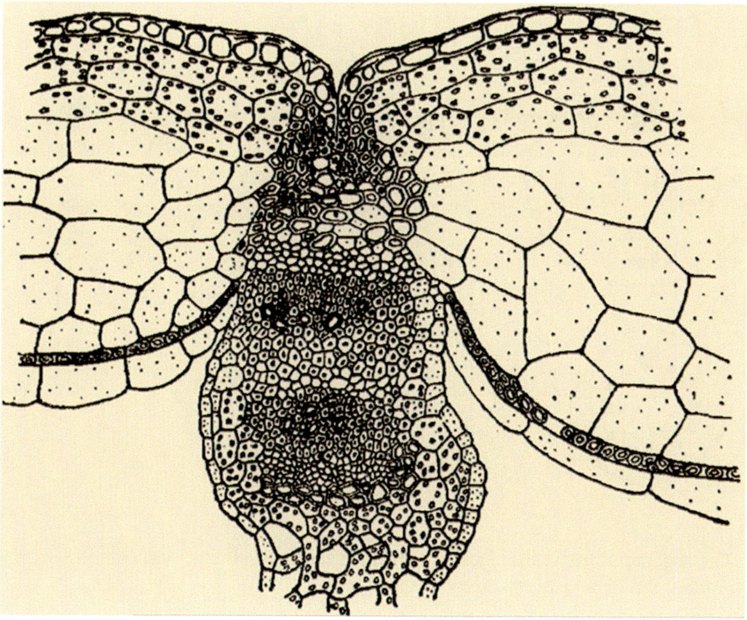

Fig. 8.6 The transection of a normal rapeseed ovary wall, showing the morphology of the outer main fruiting vascular bundle, the inner main fruiting vascular bundle, and the surrounding tissues

the lower portions of the multiple axes. The vascular bundles formed under the induction by growth hormones, which are downwardly transported from the cleft formed at the top of herb's axes, do not have meristematic cambia and thus do not have the ability to keep forming xylem and phloem. Therefore, the stems cannot keep develop new xylem and phloem to make themselves woody. If its cambia could

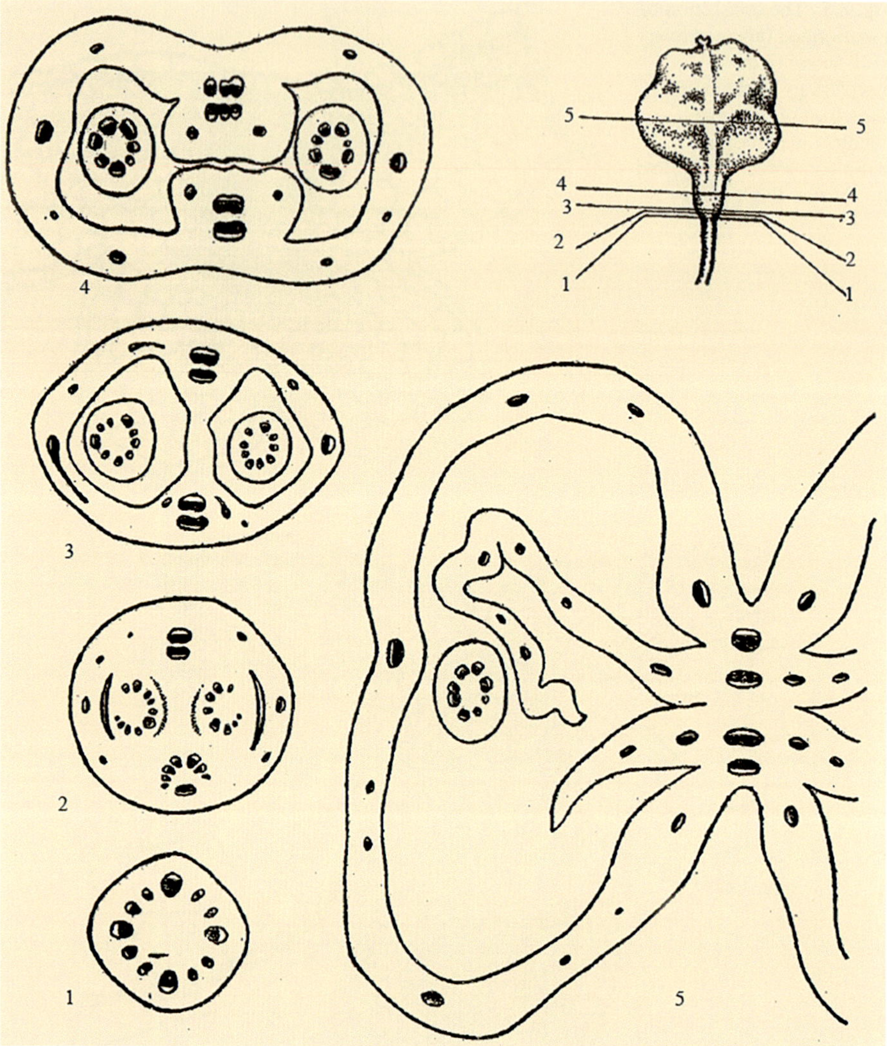

Fig. 8.7 The sequential transections of a malformed rapeseed silique, showing the distribution of vascular bundles and their relationship with the carpels

develop xylem inward and phloem outward to make the stem woody, the stem could continue thickening, in addition to the new formation of axes, which makes the stem increase its length, and thus the plant could grow into a tall tree, just like the Californian redwood we talked about earlier. Evergreen tree *Citrus grandis* (L.) Osbeck also develops in this way, so are many evergreen and deciduous trees. These trees can usually live for several decades or even hundreds or thousands of years and need 7–8 years of germination and flowering. For example, pomelo (*Citrus grandis* (L.) Osbeck) may need 20–30 years, and so is ginkgo (*Ginkgo biloba* L.), which is

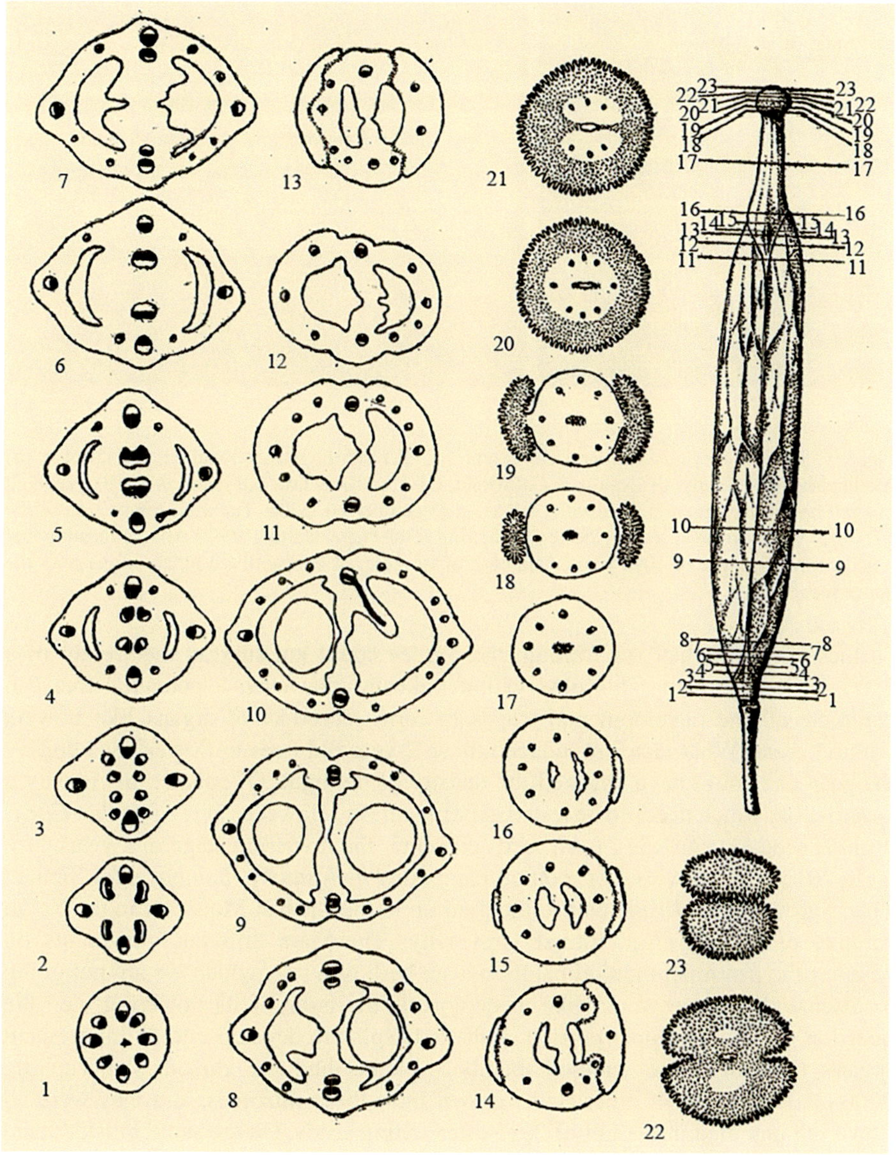

Fig. 8.8 The sequential transections of a normal rapeseed silique, showing the distribution of vascular bundles and their relationship with the carpels

called "Grandpa-Grandson Tree" (i.e., a man plants a ginkgo tree when he was young and has to wait for his grandson to eat the ginkgo fruits). However, the author, in 1945, saw that a few 2-year-old *Citrus grandis* seedlings sprouted from seeds scattered in the soil very rich in humus and shaded by a large pear tree. Complete healthy flowers had already developed on the top of these seedlings. No fruit was set

Fig. 8.9 Sequoia (*Sequoia sempervirens* (Lamb.) Endl.) and the transectional view of its stem. (**a**) A comparison of a car with sequoia in California, USA, to show how tall its trunk is; (**b**) a sequoia tree fell near Bear Creek in California, USA, on December 26, 2006. The third transection of the stem is 21 m above the base with the beginning ring at the center starting at 912 A.D., indicating that it had grown for at least 1094 years. The base of its stem, of course, lived much longer than the upper section

on these trees. If they are fruiting, these trees could not support the weight of a 1000–1500-g pomelo. This observation indicates that even a perennial tree can complete all the developmental stages to form reproductive organs like flowers within 1 year. What factor(s) induced those 2-year-old trees to flower? Shading by the pear tree could have changed the density of the light. Of course, this is only a speculation, which needs to be tested experimentally. However, it is a fact that *Citrus grandis* seedlings growing from seeds can enter the flowering stage in 2 years.

In 2015, a large school of mottled bamboo (*Phyllostachys bambusoides* Sieb. et Zucc.) grew at the hillside east to the road up to the Laoban Mountain in the Ya'an campus of Sichuan Agricultural University. There are different viewpoints on whether the perennial mottled bamboo is an herb or a tree, which we are not going to discuss. What we would like to explore is the issue of its flowering, i.e., the question of how the upper portion of the axis splits to develop into an inflorescent organ. One of these mottled bamboo plants had its rhizomes occupying 2000–2600 m^2 with 40 tiller stalks grown from these rhizomes, and they were all flowering and died in the fall of 2015 after setting seeds. On the same hillside lived many mottled bamboo plants planted at the same time. Yet these plants had thousands of tiller stems, which grew very lush, and none of them flowered. Why did this mottled bamboo plant flower but all others did not and they all grew at the same thermophase under the same sunshine condition? It seems that the difference between this flowering bamboo and all other nonflowering bamboos in the same growing area is due to some internal factors. How long can a mottled bamboo live until flowering? It can live for 20 years or more without flowering. This mottled bamboo forest was planted in 1994, about 21 years from 2015. Among them, the oldest nonflowering plant had been grown for 20 years. Most bamboos flower once in their lifetime, and they die after flowering. For mottled bamboo, which reproduces with rhizomes, it means the death of a large piece of a bamboo forest, just like in the case of annual/biannual wheat. For the mottled bamboo cluster on the Boss Mountain that flowered, the thin and small stalks developed in the early years (should be in 1995–1996), the strong and thick stalks developed after 1996, and strong and thick stalks of less than 1-year-old all flowered at the same time. It seems that the causal factor(s) for flowering is/are internal with the whole plant as the unit, not the stalk as the unit. The stalks of 20-year-olds flowered at the same time as the stalks of less than 1 year old. That all other plants established at the same time did not flower seems to indicate that bottled bamboo does not have a limitation for plant age or stalk age. Flowering is determined by the physiologic conditions of each plant. The physiologic conditions that induce flowering need to be explored experimentally. However, it must be some internal and external inducing factors for flowering.

A large mottled bamboo stalk grows from a large bamboo shoot. That means they grow large not later but in the early development stage of the bamboo shoots, when the meristems at the lower portions of the multiple axes have not yet differentiated into the vascular bundle system and the subregional meristems have already gathered into a group that vigorously grows into a large compound of internode tissues, which in turn differentiates into a large stalk. This growth model is the same as those of the columnar stalks of palm tree and gymnosperm plant cycad, which depends on the initial divisions of the internode meristem, not on the subsequent proliferation of the cambial meristem. Therefore, the large columnar stalks of mottled bamboo, palm, and cycad can no longer thicken. This is quite a different growing model from that of tree stems. Palm and cycad can flower yearly after entering the flowering stage, and their stalks continue growing up yearly. This is completely different from mottled bamboo.

The aforementioned young pomelo seedlings that bloomed in the second year did not flower again on their axillary buds developed after blooming but had gone under vegetative growth for 3 years. In 1948, the author transplanted three of them to a place near a wall and used them as rootstocks to graft scions of Shatian pomelo in 1950. These trees then flowered every year. This observation could not be continued because the garden was going to be used for other purposes in 1955. If these three pomelo trees had not been grafted, they probably could, like those pomelo trees growing from seeds, only start flowering in seven to eight years. In other words, the scions had prompted their earlier flowering.

Biannual plants, such as wheat and cabbage, die after flowering and seed setting. However, kohlrabi (the corm cabbage), after flowering and seed setting, maintains dormant axillary buds in the upper portion of the stem below the pedicel. The author did an experiment by cutting the inflorescence of a kohlrabi plant after flowering and observed the germination of the dormant buds and the formation of corm again (Fig. 8.10). This result indicates that these axillary buds returned to the first morphological developmental stage and thus suggests that morphological developmental stages are reversible in individual buds. Due to the lack of cambium development in the herbaceous stem, new stem tissues could not be regenerated for thickening and

Fig. 8.10 Photos showing the formation of new swollen stems of kohlrabi, which were formed because cutting off the inflorescence broke the dormancy of the axillary buds on the upper stem and restarted the first developmental stage. (**a**) The whole plant, showing an old corm at the bottom and a new corm above; (**b**) a new corm formed from a once dormant axillary bud due to the cutting off of the inflorescence

continuous growth. However, unlike the other kohlrabi plants, the herbaceous stem of the experimental plant did not senesce to die after flowering and seed setting but had normally lived for more than a year as the new corm grew. It seems that the regrowth of corm must have changed the inner physiological conditions of the stem to continue its life. Although the old stem did not become green again, it prolonged its life for more than 1 year. This is not the only case when the developmental stages of a stem are reversed. In *Leymus psudoracemosus* C. Yan et J. L. Yang, seedlings grow on spikes; *Polygonum viviparum* L. grows bulbils on the middle portion of its spikes; and *Chlorophytum comosum* (Thunb.) Jacques grows seedlings on its inflorescence. All these are similar phenomena of developmental stage reversion.

In summary, there are two major types of stems that are composed of multiple axes. For the first type, the meristem at the lower ends of the axes (i.e., the internode meristem) develops into vascular bundles under the inducement of growth hormones from the top. During the differentiation of xylem inward and phloem outward, a thin layer of parenchyma forms cambia around the stem for horizontal development. These parenchyma cells maintain their primary meristem status and continue the differentiation of xylem cells inward and phloem cells outward. If the cambia continue to exist and actively divide, it will keep the stem, which is formed by the lower ends of the multiple axes, and continue to thicken while elongating to become such a dazzling organ as a huge tree trunk. If the cambia are not active, thinner and shorter stems of shrubs form. For the second type of stems, the meristem at the lower portions of the multiple axes (i.e. the internode meristem) keep dividing and merge with each other to form the parenchyma in the stem. Induce by the growth hormones, these parenchyma differentiate into vascular bundles. These vascular bundles then differentiate inwards into xylem and outwards into phloem, not preserving any parenchyma between the two to be the original meristem. This type of stem, which is formed by merging the lower ends of multiple axes, has lost its ability to grow horizontally. The stems of monocotyledons, herb dicotyledons, and sago palms are formed this way.

8.2 Formation of Leaves and Flowers from Axes and the Mechanisms for Flowers and Fruits to Complete Their Productive Functions

Flowers and fruits are formed from axes, and the mechanisms for them to complete their reproductive function are still a mystery. As we have discussed many times above, in terms of morphological development, calyxes, petals, stamens, pistils, and ovules all evolve from leaf clefts. They are different in morphology and function but are genetically the same. What causes them to change differently? These changes are noteworthy in spermatophytes, particularly in angiosperms. Botanists have noticed the inducement of external factors since the nineteenth century. In 1865, J. Sachs did an experiment of shading purple morning glory and nasturtium. While the lower leaves of these plants were exposed to long daylength, shading the upper part from light induced flower formation there. When all the leaves were shaded for short daylength, no flower bud formed. Sachs stated that flower-inducing material was produced in the leaves under long daylength and transported up to the upper part to induce flower bud formation. In 1914, Tournois discovered that light controls flower formation in hops and marijuana (*Cannabis sativa* L.). In 1920, American botanists Garner and Allard discovered that tobacco (*Nicotiana tabacum* L.) cultivar Maryland Mammoth, under long daylength in summer, grew 3–4 m tall and only developed leaves but no flower. If the tobacco plants were moved into a greenhouse in the fall to prevent them from freezing to death, they could bloom and set seeds under the short daylength of the fall and winter. New plants sowed in the greenhouse in the fall grew less than 1 m tall and started to flower. This suggested that short daylength induced flowering in tobacco.

Research and observations over 100 years have told us that some plants, such as wheat, barley, cabbage, radish, Chinese cabbage, and wood-plant osmanthus, need long daylength to induce flowering, while some plants, such as rice, maize, soybean, tobacco, marijuana, and chrysanthemum, need short daylength to induce flowering, and some plants can bloom under both long and short daylengths, basically having no daylength requirement, and thus are called neutral plants. There are some special varieties that are neutral. For instance, in rose (*Rosa chinensis* Jacq.), some varieties are called "monthly flowering." They can bloom all year round and thus are neutral. An osmanthus variety, "Yueyue Gui (monthly flowering osmanthus)," is also neutral in daylength.

In 1918, German botanist Gustav Gassner discovered in his experiments that winter rye needed to be exposed to low temperature in the seedling stage to flower, but spring rye did not need to. In 1925, Trofim Denisovich Lysenko, a staff of Ukrainian Agricultural Experimental Station of the Soviet Union, treated germinating winter wheat seeds with cold temperature in spring and then sowed them in the spring. He found that the cold-treated winter wheat plants were able to head and set seed, just like the spring wheat. He called this chilling treatment jarovization. This experiment indicates that winter wheat needs a period of prolonged chilling temperature to enable the induction of flower formation by long daylength. This stage was later called the vernalizing stage, and this chilling induction is called vernalization. Spring wheat is genetically different from winter wheat. It is able to flower under long daylength without the need for inducing chilling. This research by Lysenko was for agricultural technology, not for plant physiology.

Regarding T.D. Lysenko, the author has to leave the main topic and say a few words. Some pieces of literature claimed that he did not even understand any foreign language. For such a derogatory comment, the author does not know whether it is true or not. However, Lysenko believed that the heredity of wheat, after vernalization, has changed from winter type to spring type and, thus, "acquired traits can be inherited" and that developmental stages are irreversible. These arguments clearly do not conform to objective reality. Since the twentieth century, numerous objective experiments have confirmed that external factors can only cause genetic changes if the DNA's base arrangement in the cell is changed and acquired traits are not inheritable. The aforementioned experiment, in which kohlrabi regenerates a corm of the first developmental stage on the inflorescence stalk, fully demonstrates that a developmental stage is reversible. We do not agree with the "Michurin genetics" fallacy of Lysenko. However, it exists objectively that some plants do need a thermophase or vernalization. We cannot negate the objective and real facts that Lysenko found when we reject his paradoxes.

As mentioned earlier, W.W. Garner and Allard of Beltsville, agronomists of Agricultural Experimental State of USDA in Maryland, published two papers in the Journal of Agricultural Research in 1920 and 1923, respectively, reporting their experimental observations of tobacco (*Nicotiana tabacum* L.) cultivar Maryland Mammoth. This tobacco cultivar, when grown in the open field in the summer, can grow to 3–5 m but only grow leaves, not flowers. It cannot survive the low temperature of the fall in the open field. If some plants are moved into a greenhouse, they will flower under the short daylength in the fall and winter. Besides, the greenhouse-grown plants of the same cultivar will soon flower and set seeds when the plant height reaches about 1 m. The difference between the two groups is the daylength. Maryland has more than 14-h daylength in the summer but less than 14-h daylength in the greenhouse in the fall and winter. Therefore, they concluded that Maryland Mammoth is a short-day plant. It can be made to bloom and set seeds in long daylength in the summer if artificial shading is applied. This study confirmed that flowering in plants needs physiological changes induced by photoperiod.

In 1933–1937, M.C. Cajachjan, a plant physiologist of the Soviet Union and his collaborators, C.R. Alexandrovskaja and C.R. Jarkovaja, published a series of eight papers in the Proceedings of the Academy of Science of Soviet Union to report their experiments, conclusions, and hypotheses. They conducted shading experiments on *Dendranthema morifolium* and *Perilla nankinensis* and made 14 conclusions. Generally, flowers were formed on the top of the stem when it was not bagged (>10-h daylength) but the leaves were bagged (10-h daylength); on the other hand, no flower was formed on the stem top when it was bagged but the leaves were not bagged. From these experiments, it was concluded that short daylength induced leaves to produce a hormone, which flowed to the stem top to promote the transformation of growing points at the stem tip to flowers. However, the stem top could not produce such a hormone. They named this hormone "florigen."

It has been approved by a grafting-shading experiment in cocklebur (*Xanthium aibiricum* Patrin.) that, when several stems are grafted together, as long as one of the stems has been treated by shading for short day and thus has had flower-inducing substance produced, the flower-inducing substance can flow into those stems that have not been treated to make them flowering.

It seems that the leaves did produce the flower-inducing substance in the shading treatment. However, up to this day, no one has isolated such a "florigen." What was actually obtained in the related plant physiological experiments are as follows: indole acetic acid, gibberellic acid, and the estrogen-like and androgen-like terpenoids, such as sitosterol and stigmasterol, and flowering locus T (FT) protein. They all play a role in flowering physiology. These are physiological problems, and we should let plant physiologists take care of them. What we want to know is why, in the same plant, some growing points (i.e., new axes at different places) are able to develop into inflorescence organs but others can only develop into foliage branches.

For instance, in Magnoliaceae, it is the apical buds on the stem top that change into flowers in magnolia (*Magnolia denudata* Desr.), but in white champaca (*Michelia alba* DC.), it is axillary buds that develop into flowers. Another example is peach (Prunus persica (L.) Batsch). Its apical buds and axillary buds are all leaf buds. Only the lateral buds that developed by the sides of an axillary leaf bud in the middle of a fruiting branch can evolve into flowering buds.

Flower buds are only differentiated from the lateral buds developed by the sides of an axillary bud in the middle portion of a fruit branch. How do such differences come? Some will say that "this is because they have different genetic traits" or more "scientifically" say that "their DNA is different." Yes, the base pairs in their DNA molecules are arranged differently. But how do these different base pairs in these DNAs express so as to cause the neonatal axes to form in different places of the body to induce different physiological changes that lead to different physiological states of their own? This is a question that is still unanswered! Scientifically, it is still unknown.

Also, serpent grass (*Polygonum viviparum* L.) in the Polygonaceae family, which is common on the Tibetan Plateau, develops bulbils from the growing points in the middle to the lower portion of the spike, which should differentiate into flower buds. In other words, it uses nutrient materials stored on the stem to form ball-like leaf buds, an asexual reproduction organ. We can also see this phenomenon on the spikes of *Leymus pseudoracemosus* C. Yen et J.L. Yang, which grows on the Tibetan

Plateau, and *Leymus ramosus* (Trin.) Tzvelev. On the spikes of these Leymus grasses, growing points that should differentiate into flowers have developed into small leaf buds, which, if falling into the ground, can grow roots and thus become asexually reproduced plants. Similar seedlings can also be seen in the inflorescence of spider plant (*Chlorophytum comosum* (Thunb.) Jacques). In these cases, physiological differences between the higher developmental stages are all erased, and all have returned to the first morphological development stage.

Regardless of the formation of a small plumule in seed after flowering and fertilization or the growth of a new corm on the top of the inflorescent stalk of kohlrabi, the physiological state of these axes has returned to the first morphological developmental stage. All physiological changes induced by external factors, such as thermophase and photoperiod, are released, and all the hormones induced by these external factors are also removed. Their genetic materials have not changed, and DNAs are still those DNAs and still have those base pairs with the same arrangements. These events are repeated, cycle by cycle, between generations. This, of course, is also the law of the genetic mechanism formed under evolutionary selection.

There is another question. From the aforementioned experimental results, plants must be induced by external factors to change their physiological products, which leads to changes in developmental stages and the corresponding changes in the shapes and properties of organs. For example, leaf clefts composed of axes change from cotyledons to euphyllae, from different euphyllae to different floral organs and seeds and to bulbils or seedlings for asexual reproduction. During these processes, thermophase, photoperiod, and various developmental stages all seem to be essential. But we also see that wheat has various cultivars with the thermosensitivity of strong winterness, winterness, semi-winterness, springness, or strong springness. Chilling temperature may or may not be required for the change from the first developmental stage to the second developmental stage. Chrysanthemum is a typical short-day plant, but some varieties can flower in May under long daylengths. Short-day osmanthus (*Osmanthus fragrans* Lour.) is grown widely in the Ya'an campus of Sichuan Agricultural University. It is native to the southwest region of China, growing wildly in the Qionglai Mountains near Ya'an. All the osmanthus plants are from seeds with a large range of genetic variations. The flower color varies tree by tree from white, yellow white, yellow red to dark yellow red. The flowering time also differs tree by tree, from July all the way to October. Some varieties flower all year long; therefore, they are being called "Yueyue Gui (monthly flowering osmanthus)." It seems that the inducement of flowering by long or short daylength is also dispensable.

Woody angiosperms, no matter which family, genus, or species they belong, all have two kinds of branches, i.e., the long branch that grows vigorously and the short tiny lateral branch that does not grow vigorously. Both branches grow under the same thermophase and photoperiod. However, the meristem on the vigorously growing long branch in most cases does not differentiate into flowers. Flower differentiation most often occurs on the small lateral branches, which do not vigorously grow. Someone calls the former an excessively growing branch and the latter a short fruit branch. In Sichuan, there is a chrysanthemum cultivar called "Green Jade" or "Green Mum." The cells in its petals contain chloroplasts, and thus the flower displays green-white color. This cultivar is very sensitive to nitrogenous fertilizers. A little bit too much nitrogen will cause it not to flower but keep it in a vegetative growth state having only green leaves. If a branch in this vegetative growth state is harvested for asexual cottage propagation in soil poor in nitrogen, it will produce green-white flowers in the fall. In these two cases, C/N ratio has a greater effect on flowering than thermophase and photoperiod.

In lower plants, such as hornworm (*Anthoceros laevis* L.), a sperm produced in antheridia on a haploid green thallus of the sexual generation fertilizes an egg in an archegonium to form the cell of the diploid generation, which grows to a cell cluster by mitosis. The cell cluster expands bipolarly with one pole merging into the bottom of the archegonium to absorb water and nutrients supplied by the thallus and the other pole growing upward to become an axial structure (an axis). Cells inside the upper portion of the axis undergo meiosis to form numerous haploid tetraspores. The upper portion of the axis, therefore, becomes a sporangium. Mature sporangium splits into two pieces, releasing the haploid sexual spores, which then develops into a thallus of the next generation. This is the life cycle of hornworm. The bipolar growing axial structure of its asexual generation is the primitive axis of a plant.

Hornworm's capsula grows vegetatively and the capsular petals become leaf clefts specializing for photosynthetic nutrition, establishing the primitive form of higher plants. New axes increase within the leaf clefts to become various organs of higher plants, which evolve while surviving selections; external factors induce internal physiological changes; different yet interconnected developmental stages have gradually formed. From the germination of a spore or seed to the reproduction of spores or seeds, life cycles repeat generation after generation. Under different physiological conditions, genetically identical axes develop into organs of different shapes and functions to complete the stage cycle of life. Plants can live endlessly generation after generation, although at present, humans have not clearly understood the physiological mechanisms of different plant organs and the need to study them further, just as we can only ask questions as we did above. Nevertheless, development of phylome from the leaf clefts on the upper portion of the axes has undergone a graduate evolution. In angiosperm, it has evolved into higher forms, such as cotyledon, basal leaf, various cauline leaves, calyx, petal, stamen, pistil, integument and nucella. Each has its own distinguished shape and unique function to complete a life cycle. These variations have survived the natural selection due to their fitness. The genome, which consists of the preserved DNA mutations, will orderly express under the external eco-environment, enabling this species (genome) to reproduce. What is elaborated here is how some of such higher plants living in the world have their organs constructed.

Agricultural production is the production of certain higher plant organs. Improving the quality and quantity of such organs to make them more suitable for the purpose of production by humans is what plant breeding is for. More specifically, from the viewpoint of scientific research, breeding is economic phytoteratology. If we do not understand the basic units of the organogenesis of these higher plants and their evolution, we will be somewhat blind or, in other words, "taking a chance" when drawing the breeding roadmap. Taking a chance will not work. In order to obtain expected results, we must practice according to objective laws and apply science and technology. The scientific knowledge we have now is capable of guiding our work. Those unknown natural laws need to be explored. Only in this way can our work be further advanced. Scientific knowledge comes from comparisons and analyses of practice, not from illusory delusion. There is still a lot of unknown scientific knowledge to learn.

8.3 Ideotype: Designing Organs for Novel Cultivars

Crop cultivars are selected from natural variations to meet human needs, not for the needs by the plants themselves. In other words, the selections are not beneficial to plants themselves. Taking the aforementioned Chinese cabbage and napa cabbage as the examples, people love to eat their tender albino inner leaves which are due to lack of sunlight because of their tight entanglement. However, if people would not cut the entangled leaves open, these plants would not be able to flower. These entangled leaves, thus, are the malformation that is not good to the plants themselves. Therefore, the vast majority of crop cultivars are malformed plants. Of course, there are human needs that are consistent with the needs of the plant itself. Even such a plant has organs and thus organogenesis that are different from the original wild type, and therefore, they are different from the original wild plants. Most of the human demands from crops are for their individual organs, not for all their organs. For instance, what we need are the roots of radish, the tubular stem of potato, leaves of cabbages, abnormal inflorescence of cauliflower, the pericarp of peaches and plums, the seeds of sesame and peanuts, and the pericarp and seeds of wheat and rice. Comparing to the original, these are all abnormal, only difference here is whether the variation from the original is great or small. Therefore, plant breeding can be regarded as a science of economical abnormity of plants which studies abnormal plant organs.

Austria's C.M. Donald (1968a, b) introduced the breeding term "ideotype." He presented a paper entitled "The Design of a Wheat Ideotype" at the Third International Wheat Genetics Symposium held in Canberra, Australia, in 1968. In the same year, he also published an 18-page long paper entitled "The Breeding of Crop Ideotypes" in Euphytica to explain his "ideotype" concept. The title of his Wheat Genetics Symposium paper, "The design of a wheat ideotype", is very good, because "design" is to draw a technical plan according to human economical needs. In other words, breeding is optimizing a crop's organ(s) or designing ideal organ(s). What are his thoughts on the design of wheat organs? He said that the ideal type of a field crop, no matter what species it is, should have some common characteristics, which are as follows:

1. A field crop is a population of a single type, which is different from the population of a natural plant or forage crop, which is composed of multiple species.
2. All the plants in a field crop population have the same or a similar genotype, which makes competition among the individuals of the crop population different from those between different genotypes or species.
3. Due to the strong competition between plants within a crop population for things needed for growth, the yield per plant in the population is lower (10–20% or even lower) than a single plant if it is cultivated alone.
4. In modern agriculture, heavy fertilization surely results in a high yield.
5. If the growing conditions for a crop have all been improved, the growth of the crop will depend heavily on its maximum light use efficiency. If rainfall is abundant or it is irrigated, higher yield of the crop is decided by how it uses the light.
6. Except of forage crops, a crop generally is only important for some of its organs. People who grow a crop are not interested in how much dry matter they can get from it but in what product they can get out of it and how much economic value the product has.

The above descriptions and the design of wheat ideotype (Fig 8.11) described in his Wheat Genetic Symposium paper indicate that Donald was correct in emphasizing genetic uniformity or similarity in a crop population. No matter how strong the tillering heritability a cultivar has, a high-yielding plant in a normal population of this cultivar usually has only one normal spike plus a small, less developed spike. From the viewpoint of yield, it means one-and-a-half spikes. Single culm is a favorable architecture, and many gramineous crops have a malformed single culm. Whether millet, sorghum, or corn, good cultivars all have a single culm. This is a random result of long-time unconscious artificial selection. Donald stated in his design drawing that there should be "a high ratio of seed roots," which is obviously wrong and is not consistent with the basics of plant organ structures. The number of seed roots is fixed in wheat. As one can see from Fig. 7.18, it can only have six. All other roots are grown from tiller nodes. Even these six "seed roots" are secondary roots growing from the coleoptile nodes of caryopsis. The true seed root in wheat, i.e., the low portion of the first axis, has already evolved into coleoptile and thus lost its function as a root. Other gramineous crops that have a single culm all grow a large number of supporting roots from nodes above the tiller nodes near the ground surface, which wheat usually does not have. Without the supporting roots, lodging will occur due to the heavier culm and large spike of the single culm. Therefore, when people unconsciously selected high-yield, single-culm varieties of maize, sorghum, and millet, they selected the supporting roots. Nevertheless, the importance of the support roots was not recognized by Donald.

The author has conducted decades of research on single-culm wheat and, from 1952 to 1960 at the earliest, did an 8-year continuous observation of the single-culm wheat and barley plants made by artificially removing the tillers. The results were published in a paper entitled "The Study of Tiller Removal I: The Effects of Removing Tillers from Wheat and Barley in Successive Generations" in 1963 in

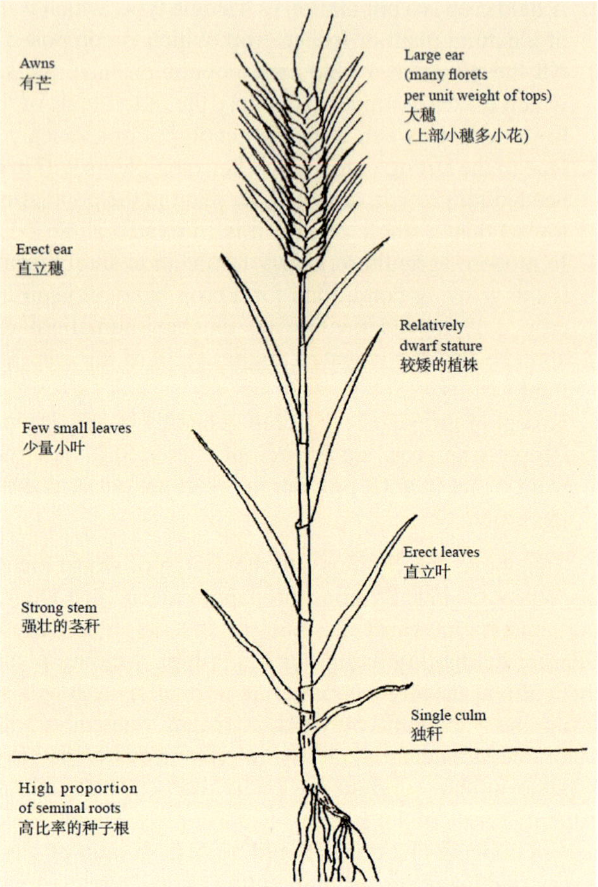

Fig. 8.11 The design of wheat ideotype. (Adapted from Figure 1 in Donald 1968b)

The Genetics Colloquium, Issue 5. The experimental results showed that, at the same time as the formation of the single culm, the dosages of indole acetic acid, gibberellic acid, and the like changed, which led to changes in organ structures. Removing tillers from wheat and barley plants have caused significant changes in economically important traits such as jointing time, plant height, the color, width and length of leaves, the culm thickness, spike shape, TSM (thousand seed weight) and crude protein content in seeds. Especially, the leaves became long and wide and onefold wider and longer than normal leaves. The leaf color was dark green. Although seeds became larger and TSW increased, the filling was poor, the starch content was low, and the seeds were not plump, which caused the illusion of an increase in crude protein content. When plants from the seeds of plants that have gone through several generations of artificial tiller removal were no longer subjected to tiller removal, all of the changes disappeared, indicating they are not inheritable.

Referencing the examples of other single-culm gramineous crops, it is possible to select a wheat genotype that has support roots, a single culm, smaller leaves, and

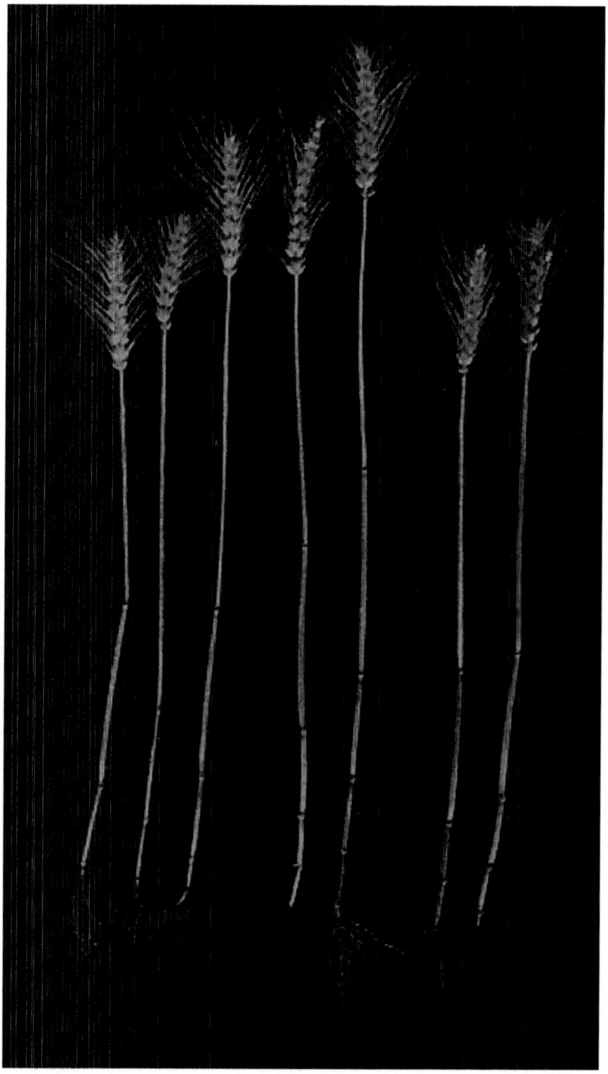

Fig. 8.12 Single-culm wheat plants

good grain filling. Since the 1960s, the author has paid attention to breeding such genotype and has also done some experiments on the ideotype of wheat. Mutant populations of single-culm wheat were obtained in the 1970s (Fig. 8.12). Two kinds of stably inheritable large spikes (branching spike and supernumeric spikelets) were also obtained successively in the 1950s and the 1980s. Multiyear observations and studies of these materials were conducted in the meantime. Although single-culm mutants were obtained, the problem of small leaves has not yet been solved. The single-culm mutant plants are similar to those that were treated with tiller removal and also had high indole acetic acid and gibberellic acid, which caused strong vegetative growth with long and wide leaves and poor grain filling and thus shrunken gains. In addition, although supporting roots appeared on individual plants, the number was too small and not enough to support the single-culm plant from lodging. With reference to other gramineous plants, it is possible to breed single-culm wheat cultivars with a large number of support roots, smaller leaves, and normal grain filling. The problem of seed weight has been solved since materials with over 50-g TSW have been obtained. Studies on seed number have also been conducted. In 1965, a paper entitled "On Developmental Genetics of Spike Structure of Common Wheat" was published on The Genetics Colloquium, Issue 6. This paper summarized the studies conducted between 1956 and 1963 on some common wheat organs relevant to increasing seed number and made the following conclusions:

1. Any wheat can exhibit branching spikes if all the required conditions are met; branching is not controlled by the special recessive branching gene *bh*; wheat lines with different heredities may require different conditions for branching.
2. The relationship between heredity and trait expression can be summarized as follows:

   (a) Different heredities may express the same phenotype in different environments.
   (b) The same heredity may express different phenotypes in different environments.
   (c) Having the same phenotype does not necessarily mean having the same heredity.
   (d) Phenotypes are different, but heredity could be the same.

3. Branching spike, supernumerary spikelets, and high seed number per spikelet are all formed by a prolonged third developmental stage. In addition to differences in the length of the developmental stage, the growth rate of the second spike axis is also an important determinant. Although the materials studied differ greatly in morphology, the differences were very small in terms of developmental genetics.
4. Researchers who breed branching wheat should not look for a *bh* gene carrier but should select those that enter the third developmental stage earlier but could not enter the fourth developmental stage under local cultivation conditions.
5. Materials with branching spikes are ecotypes that have a particular longer third developmental stage under local cultivation conditions. Using them as parent materials, it is possible to obtain various types of large spikes with a high floret/seed number, such as branching spikes, unbranching but having supernumerary spikelets or spikelets with a high floret/seed number, etc.

Following the breeding plan for a long third developmental stage, the author bred, from wheat cultivar Liying 3, varieties with stably inherited branching spikes, which represent the materials with a long third developmental stage. In other words, their genes express the traits of a long third developmental stage under the cultivation conditions in Sichuan. Crossing it to Fengshou (a semi-spring cultivar from Albania), materials with supernumerary spikelets were obtained. The number of their spikelets per spike can reach 28–32, about 6–10 more than the usual wheat spikes

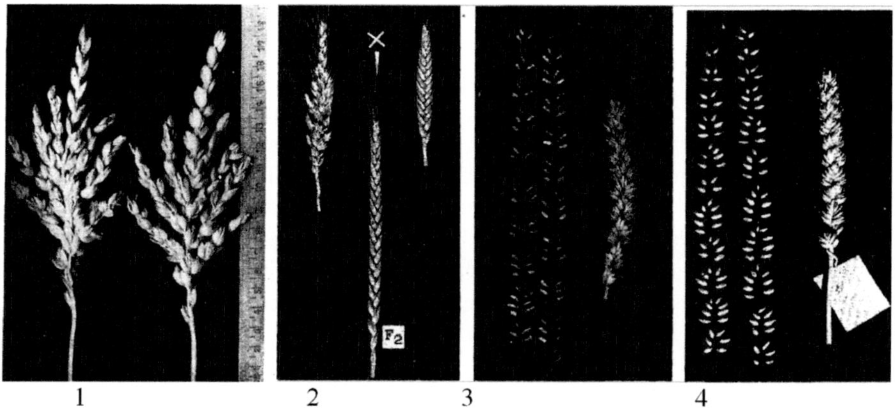

Fig. 8.13 The spike architectures of hexaploidy common wheat with a prolonged third developmental stage. (1) The spikes of Branching Chengdu Guangtou; (2) a supernumerary F_2 spike and its parents Branching Liying 3 and Fengshou; (3) a (Branching Chengdu Guangtou × (Nanda 2419 × Zhongnong 483)F2) F_4 spike with a high seed number per spikelet; the seeds are arranged according to their positions on the spike; (4) a ((9× − 1 × 51 Mai) F_1 × (Branching Zhongnong 483 × Branching Chengdu Guangtou)) F_1 with a high seed number per spikelet; the seeds are arranged based on their positions on the spike

(Fig. 8.13(2)). The author also selected several varieties with stably inherited branching spikes from cultivars Chengduguangtou (aka, Chinese Spring) (Fig. 8.13(1)), Zhongnong 483, and Zhongnong 28. Also obtained were varieties with high floret/seed number, which increases seed number per spikelet from 3–4 seeds to 8–9 seeds (Fig. 8.13(3), (4)).

After nearly 40 years of observation and test, the author believes that the possibility to obtain a high-yield plant type with increased seed number by using branching spikes and high floral/seed number is very small. Henan province of China has a tetraploid wheat (*Triticum turgidum*) landrace "Foshou", which has branching-spike. It has a stably inherited branching spike, large spikes, and a large number of spikelets and seeds, but the seeds are small and the yield is not high. The branching common wheat bred by the author also showed large spikes and a large number of spikelets and thus seeds per spike, but again, seeds are small. What increased are glumes, rachides, and rachillae. The approach to increase florets per spikelet is also feasible. Seed weight may increase, but the grain size of the lower and upper spikelets differ greatly. Therefore, the smaller grains produce excessive bran after milling, leading to a reduced value of the product.

The ideal wheat spike architecture should be unbranched, like the rye spike architecture; it should have a high number of spikelets, but each spikelet only sets two seeds. In this way, it can effectively set many large, heavy, and uniformed seeds. The author has noticed that there are some Tibetan landraces that only set two seeds per spikelet. Unfortunately, they also do not have many spikelets, and thus the total seeds and the seed weight per spike are not increased.

Based on observations by the author in the spring wheat region in Sichuan, the ideotype of wheat there should have a single culm; small, near-erect leaves; 28–32 spikelets; each spikelet only setting seeds in the lowest two florets; 50–60-g TSW; genes for high-quality proteins; and high resistance to powdery mildew and the three rust diseases. In the winter wheat region of the yellow river valley, freezing tolerance during the seedling stage and thus strong winterness are needed. Therefore, more research is needed to find out whether the single culm is the ideotype for this region or not.

Studies have revealed that the number of seeds per spikelet is controlled by incomplete dominant genes on wheat chromosomes 6A, 1B, 4B, 6B, 2D, 4D, and 4A or 5A. The study by Youliang Zheng (1992) suggested that chromosomes 1B, 6B, and 4D of the wheat lines studied carry previously unknown new genes that control seeds per spikelet. Therefore, the germplasm for two seeds per spikelet does exist in the spring wheat region in Sichuan, but it is not clear which gene(s) control (s) it. Nevertheless, this does not prevent breeding for cultivars with two seeds per spikelet.

Many of the grass crops, such as maize, sorghum, and millet, have a single culm. Tillering is suppressed in these crops but can frequently occur in some cultivars. All their wild relatives have tillers. However, none of the single-culm cultivars of these crops have excessive dosages of gibberellic acid and indole acetic acid, which leads to increased leaf blade, poor grain filling, and shriveled grains, as seen in single-culm wheat. They also have a large number of support roots to prevent the single culm from lodging. These examples enlighten positive implications for the breeding prospects of single-culm wheat.

Taking maize as an example, it has had the greatest changes in organ structures among the cultivated crops. Therefore, maize is the model for breeding wheat ideotypes. Annual teosinte *Z. mays* L. ssp. *parviglumis* H.H. Iltis et Doebley should be the wild type of maize (Fig. 8.14), and perennial diploid teosinte *Z. diploperennis* H.H. Iltis, Doebley et R. Guzmän is a relative species. Genetically, maize, *Z. parviglumis*, and *Z. diploperennis* all have ten pairs of chromosomes with the same karyotype. F_1 hybrids between them have ten pairs of bivalents, although some are not ring but rod bivalents with only one arm connecting each other. These observations suggest that differences between the chromosomes of maize and the teosintes are only those between subtypes. They share the same genome, and they belong to the same species. Obviously, it is incorrect to call the perennial teosinte *Z. diploperennis* H.H. Iltis, Doebley et R. Guzmän as a species, and it is also arguable to designate *Zea mays* subsp. *parviglumis* H.H. Iltis et Doebley and *Zea mays* subsp. *mexicana* (Schrad.) H.H. Iltis as subspecies. Nonetheless, here we are only discussing organogenesis, not taxonomy, so we only point out the problem here without further discussion on this topic.

The greatest difference between maize and teosintes is their female inflorescences. Maize's female panicle consists of paired pistillate spikelets on the spike branches. The spike branches grow densely in (4~) 8–18 (~30) rows to form one rachis, forming a rod-shaped inflorescence. The closest maize cultivar to teosinte is "Sihangnuo (Four-row waxy maize)" found in Xishuangbanna, Yunnan, China, which has only four rows of pistillate flowers (Fig. 8.15). In other words, on each

Fig. 8.14 *Zea mays* var. *parviglumis* (H.H. Iltis et Doebley) C. Yen. (Photographed by Yang Yen)

node of the solitary teosinte spikelet, the two florets that alternatively grow on the upper and the lower rachilla nodes are aggregated and developed into a pair of florets; each spikelet then forms a pair of caryopses, which, together with the counterparts of the other spikelet, form the four rows on the rachis. The glume, lemma, and palea of each spikelet have all degenerated into short membranes, making the caryopsis naked. The internodes of the lateral branches underneath the

Fig. 8.15 Maize landrace Sihangnuo. (**a**) Lateral female inflorescence, showing the appearance of husks wrapping; (**b**) a comparison of ears between Sihangnuo maize and ordinary maize (husks have been removed); (**c**) the transections of the ears of Sihangnuo and ordinary maize, showing the number of rows

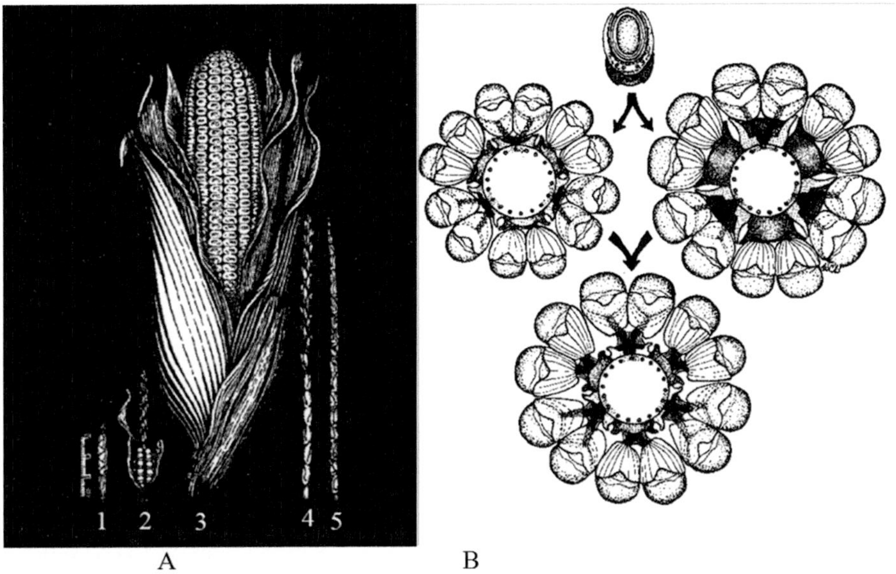

Fig. 8.16 Ears of maize and its relatives (**a**) and the two routes of domestication of maize illustrated by their transections (**b**). (**a**) Ears of maize and its relatives: 1. teosinte, 2: the oldest maize ear (about 7000 years old) found in Tehuacan of Mexico, 3: ear of modern dent maize, 4: *Tripsacum*, 5: *Manisuris*; (**b**) transections showing the two domestication routes from teosinte (top): glume degeneration (left), most seen in South America; peduncle elongation (right), most seen in North America; and both peduncle degeneration and glume elongation (bottom), most seen on maize ears with multiple ranks. (Adapted from Galinat 1971)

female panicle are aggregated and shortened to form the short peduncle below the female panicle. Leaf blades on the short peduncle are degenerated; the leaf sheaths enclose the female inflorescence to become the husk structure unique to maize. Teosintes have a large number of tillers, maybe up to 80–90 per plant. The leaves and the apically grown male inflorescences of teosintes are very much like those of maize. The female inflorescences grow on the lateral branches developed at the middle of the culm. Like the oldest maize cob of 7000-year-old found in Tehuacan, Mexico, Sihangnuo (Fig. 8.16a(2)), as described above, has the branch internodes underneath the female florescence shrunk to a short peduncle, and the leaf blade on the peduncle becomes the husks and encloses the naked caryopses on the ear. In this way, a special malformed maize cob is formed.

Sihangnuo is probably a primary maize cultivar that originated in the Americas in ancient times and spread to the southern Yunnan of China via the Pacific Islands and Southeast Asia. It has long disappeared in the Americas. This fits the model that a primary type of crop always appears on the edge of its spreading route, and it is also consistent with the discovery of husked hexaploidy wheat in Tibet and Yunnan. A variety starts to spread outwards after it was born. When it reaches the edge of its spreading zone, it has been already replaced by new variety at its birth place. As a result, an old variety usually exists only at the edge of its spreading zone. Some people think that maize originated in Yunnan because of the discovery of Sihangnuo in the region; others believe that Yunnan and Tibet are the birthplaces of hexaploidy wheat due to the discovery of husked hexaploidy wheat there. Obviously, they all are wrong because there is no wild relative species of maize in Yunnan, and Tibet does not have the parental species of hexaploidy wheat.

The multitillered teosinte mutated into single-culm maize, which does not make the leaves abnormally large and the grains fully filled. This indicates that a single-culm wheat variation with normal organ morphology and physiology can be obtained as well.

The basic inflorescence architecture of teosintes and maize is a panicle with multiple branches. The male and female inflorescences have the same basic structure. Female florets can occasionally grow at the lower portion of a male inflorescence, producing a few kernels, while a short male inflorescence sometimes grows from the apex of a maize ear. A few branched female inflorescences can often grow at the base or apex of a female inflorescence (maize ear). All these abnormalities illustrate their panicle nature. Maize ear has evolved from the basic four rows to multiple rows, which, in fact, is due to webbing and the aggregation of multiple primary branches of the female panicle. That is why maize's caryopsis ear can have rebranched malformations at the base and apex. Normal maize is a malformed teosinte with very high economic value. The construction of a multirow maize ear is based on webbing and the aggregation of multiple branches of teosinte panicle. The basis of its row of caryopses is four, i.e., one branch. The normal architecture is of 4, 8, 12, 16, 20, 24, 28, or 32 rows. Other row numbers should be caused by the degeneration of florets or the two rows of florets on one side of a spikelet. This is because a panicle branch has two rows of spikelets, and each spikelet has two fertile florets. A gramineous leaf, as described above, is constructed from multiple axes, and a spikelet is composed of multiple axes. Caryopsis is no exception. Secondary axes are the fundamental units for constructing various organs. Enzymes produced under the guidance of the information stored in DNA sequences regulate physiological mechanisms to transform secondary axes into various organs. It is an important genetic study to simply attribute various organs to genetically paired genes, but it is impossible to understand the organs themselves and their evolutionary mechanisms and pathways without understanding the construction mechanism of the secondary axes. This is knowledge of another level. As mentioned earlier, it is obviously incorrect to simply attribute the branching spikes of wheat to a trait controlled by a pair of *bh* recessive genes. It is an objective fact that wheat cultivars with different genes can form similar branching spikes under different external conditions. Only based on scientific knowledge of developmental physiology and developmental morphology can we clearly understand this objectively existing true phenomenon. Genetic analysis was a novel way in the twentieth century to gain scientific knowledge, which was lacking in the biological sciences of the nineteenth century. It is an important component of the developing biological sciences. Scientific knowledge of developmental physiology and developmental morphology is also indispensable for breeding, an applied discipline of technology design.

References

Donald CM(1968a) The design of a wheat ideotype. In: Proceedings of third international wheat genetics symposium, Canberra, Australia, pp 377–387

Donald CM (1968b) The breeding of ideotype. Euphytica 17(2):385–403

Galinat WC (1971) The origin of maize. Annu Rev Genet 5:447–478

Kuperman FM (1953) The biological foundation of wheat cultivation (Chinese translation by Zhenao Xia and Kelan Cai published by Science Press, Beijing)

Yen C (1959) On a new view of carpel morphology in *Brassica*. Acta Bot Sin 8(4):271–277

Yen C, Ngao D (1963) Observation on developmental morphology of shoot apex in wheat. Acta Agron Sin 2:411–426

Zheng Y, Yen C, Yang J (1992) Effect of chromosomes on the number of grains per spike in common wheat. J Sichuan Agric Univ 10(2):210–214

Chapter 9
Conclusions

Based on the aforementioned objective examples, the author summarizes his multiple secondary axis theory for the constitution of the organs of higher plants as follows:

1. The organs of higher plants have evolved from regenerated, bipolarly extended secondary axes. The upper portion of such an axis expands to conduct photosynthesis. It becomes hollow first and then splits to become a leaf cleft, the extended light-receiving surface. Therefore, a leaf is formed by the splitting and expansion of the upper portion of the axis. The unsplit lower portion of the first axis extends downward to become the radicle, which merges with the lower portions of the secondary axes above it to form the stem. The stem is constructed by continuous merges of the lower portion of the axis above. It is, therefore, a secondary structure. An axis grows bipolarly. It is the meristems at their low ends that make the connection with the axis below. The lower point of the secondary axis is in the status of division and, therefore, is the growing point at the lower end, in other words, the intercalary meristem which is commonly seen in the higher plants. The widespread existence of intercalary meristems in higher plants also proves that the stems of the higher plants are not originally a monolithic structure but a composition of the lower portions of multiple secondary axes. The lower portion of the first secondary axis becomes the hypocotyl, and its lower growth point constitutes the root tip.
2. Splitting at the upper portion of an axis can be simple, as seen in ferns and the monocotyledons of the seed plants; dichotomous, as seen in clubmosses, spikemosses, ginkgo, cypress, ephedra, *Welwitschia*, and the dicotyledons of the seed plants; or multiple, as seen in horsetail and pines.
3. Secondary axes grow on a leaf cleft to become a compound leaf. A compound leaf can be integrifolius; it can also be pinnate or palmate. A newly formed secondary axis on a compound leaf can develop into a bulbil or a branch (as seen in tomato), but the petiole at its base is usually a split axis (the leaf cleft of the lowest axis). In some special compound leaves, the petioles are composed of

C. Yen, *Constitution of Organs of the Higher Plants*,
https://doi.org/10.1007/978-981-19-1685-4_9

secondary axes grown on the leaf clefts; it can be an unsplit (the lower portion of the axis) axis, such as the petioles of kidney bean and pea.

4. Simple axes develop into an alternate phyllotaxy: simple axes form an alternate phyllotaxy without internode elongation; it can also be said that two leaves are aggregated by internode degeneration, which leads to the disappearance of the internode and thus the formation of the secondary opposite phyllotaxy, for example, the occasional appearance of opposite phyllotaxies on the vines of Chinese yam.

   Dichotomously split axes develop into an opposite phyllotaxy: dichotomously split axes form an opposite phyllotaxy through uneven growth and the development of leaf clefts. One leaf cleft matures earlier and stops elongation, while the other is still growing and elongating, resulting in the rise of the latter and the formation of an alternate phyllotaxy. The continuous uneven growths between the upper and lower nodes at all levels of the stem result in area differences due to the different sizes of the leaf clefts. If the angle between two areas is larger than 90° but smaller than 180°, the upper and the lower clefts will show rotation on the vertical axis. If the first-maturing cleft of the upper node lies at the left of the first-maturing cleft of the lower node, a left-handed alternate phyllotaxy will form, and if it lies at the right, a right-handed alternate phyllotaxy will form. In this way, a phyllotaxy can evolve into a left-handed or right-handed one. It can also evolve into a cyclic phyllotaxy if the internodes of the rotating alternate phyllotaxy disappear. If the angle between two areas reaches 180°, the leaf clefts between the upper and lower nodes will become para-alternate, as seen in broad bean and pea. An opposite phyllotaxy can also change into an alternate phyllotaxy if one of the paired leaves disappears, as seen in chinar.

   Multiple-split axes form cyclic phyllotaxies, as seen in horsetail and the cotyledon and multi-split leaves of pine.

5. Webbing and aggregation between axes are two natural ways of growth. They are not the growth modes of the fabricated "telomes" of the telome theory but those between axes. The formation of common cyclic floral organs and compound leaves are all constructed by these growth modes. Rotating alternate leaves can converge into a cycle of five or six leaves, which then evolve into pentaphyllum or hexaphyllum flowers. Dichotomous opposite leaf clefts can develop into four-petal flowers by the aggregation of the leaf clefts of the upper and lower axes. Webbing between leaf clefts can make a choripetalous flower gamopetalous.
6. Vascular bundles are secondary conducting and supporting structures induced by physiologically active substances such as auxin and, therefore, are secondary tissues. The vascular bundles of each axis are interconnected to form the stele of a stem and the branches and the vein system in leaves. Although this system appears as a unified entirety, it is not primary but secondary. In fact, it is a secondary entirety formed by the interconnection of the secondary vascular bundles of the interconnected axes. Whether it is a stem, compound leaf, or taproot system, they are all formed this way.

7. A stem is a union of interconnected lower portions of secondary axes; it is a secondary structure. Because an axis grows bipolarly, although its lower ends has already connected with the axis below, it maintains its primary apical growth in a period of time, forming a structure commonly called intercalary meristem.

   Because of the lag of vascular bundle differentiation, the interstitial tissues and intercalary meristem of the multiple axes are aggregated to form large, actively dividing interstitial tissue masses, which develop into a columnar stem. The columnar stems of the spinulose tree fern of Pteridophyta, cycad of Gymnosperm, and palm of Monocotyledonae are all constructed in this way, and so are the tall, thick bamboo stalks of Poaceae.

   Pine, fir, cypress, and ginkgo of Gymnosperm and trees and shrubs of Dicotyledoneae evolve a vascular bundle ring in their stems, forming xylem inward and phloem outward from the cambia evolved from the intercalary meristems. Because the cambia remain in a meristematic state for a long time, new xylem and phloem are constantly produced and accumulate to form huge trees. For instance, *Sequoia sempervirens* (Lamb.) Endl. of California, USA, can reach more than 100 m in height with a truck of 4–5 m in diameter.
8. The saliency of the leaf clefts and the segmental nature of the stem have concealed their axial origin, which led to the birth of the wrong phytonic theory.
9. The axial structure of the diploid asexual generation of primitive masses and the structural entirety of the secondary stem of higher plants, which is formed by inter-aggregation of multiple secondary axes, have concealed the segmentation that their stems originally had. This plus the "primitiveness" illusion of the degeneration and disappearance of the leaf clefts of *Rhynia major* Kidston & Lang and the misunderstanding of the stem's entirety all lead to the formation of the wrong opinions of the telome theory.
10. Different plant organs composed of multiple axes have different shapes and structures of their own, and they also have their own different functions. They have different orderly progressive stages of morphological development on the plant body, from generation to generation, through the reproductive cycles. These orderly progressive morphological development stages are formed under the induction of external factors (different temperatures, daylengths, or nutrients) and changes of internal biochemical materials. Although deoxyribonucleic acid (DNA), the hereditary materials, in their cells has not changed, the expressions are different due to the physiological changes induced by the external and internal factors. Each individual develops its organs orderly through morphological developmental stages and completes its life cycle. Each individual development starts at the first morphological developmental stage and then orderly enters various higher stages, from the formation of various nutrient organs to the formation of reproductive organs. Re-produced sexual spores combine to form a new asexual generation, which produces a seedling, seed, bulbil, or plantlet. Completing these processes, life returns to the first morphological developmental stage, disarming the physiological mechanisms of various high-level morphological developmental stages. However, to date, scientific research studies have not fully revealed the details of the physiological

mechanisms of the organ construction at each development stage of higher plants. Further research is needed.

11. In summary of the organ constitutions of higher plants, a leaf is formed by splitting the upper portion of the axis. This includes the compound leaf formed by webbing the leaf clefts of the secondary axes regenerated on a leaf cleft. The leaf cleft formed by splitting the upper portion of an axis regenerates a secondary axis within its axial lumen. While the upper portion of this secondary axis splits to form upper leaf clefts, the lower portion of the axis unites with the lower portion of the axis below to form a stem. New leaf clefts and stems come from the secondary axes formed in the split lumen of primitive axes. A root is formed through the growth and extension of the meristem at the lower end of the axis. A branch root consists of an unsplit secondary axis developed by the outward growth of the stele formed axially from secondary vascular bundles. Floral organs, including fruits and seeds of the next generation, and the asexually propagating bulbils and the equivalent vegetative seedlings for propagation, are all the heteromorphs of secondary axes due to certain physiological changes. The physiological mechanisms that induce such hetero-morphogenesis are the research areas of developmental physiology.
12. Different plant organs have different economic values for humans. An agricultural product is not all organs of the crop but a specific one. Therefore, different malformed organs produce different economic values. Malformations are formed by mutations in the four base pairs of DNA, the genetic material. Humans artificially select, cultivate, and retain mutants due to their own economic needs, forming cultivars with artificially selected malformed organs. As a technical discipline, the most important theoretical basis of crop breeding is genetics + economic plant tetratology.

GPSR Compliance

The European Union's (EU) General Product Safety Regulation (GPSR) is a set of rules that requires consumer products to be safe and our obligations to ensure this.

If you have any concerns about our products, you can contact us on ProductSafety@springernature.com

In case Publisher is established outside the EU, the EU authorized representative is:

Springer Nature Customer Service Center GmbH
Europaplatz 3
69115 Heidelberg, Germany

Batch number: 10756376

Printed by Printforce, the Netherlands